Student Study Guide with Selected Solutions

JOSEPH BOYLE

Miami-Dade College

Sixth Edition

PHYSICS

GIANCOLI

Volume 1

Upper Saddle River, NJ 07458

Associate Editor: Christian Botting
Senior Editor: Erik Fahlgren
Editor-in-Chief, Science: John Challice
Vice President of Production & Manufacturing: David W. Riccardi
Executive Managing Editor: Kathleen Schiaparelli
Assistant Managing Editor: Becca Richter
Production Editor: Dana Dunn
Supplement Cover Manager: Paul Gourhan
Supplement Cover Designer: Joanne Alexandris
Manufacturing Buyer: Ilene Kahn
Cover Photo: Art Wolfe/Getty Images, Inc.

Pearson Prentice Hall
Pearson Education, Inc.
Upper Saddle River, NJ 07458

Printed in the United States of America

10 9 8 7 6 5

ISBN 0-13-035239-X

Pearson Education Ltd., *London*
Pearson Education Australia Pty. Ltd., *Sydney*
Pearson Education Singapore, Pte. Ltd.
Pearson Education North Asia Ltd., *Hong Kong*
Pearson Education Canada, Inc., *Toronto*
Pearson Educación de Mexico, S.A. de C.V.
Pearson Education—Japan, *Tokyo*
Pearson Education Malaysia, Pte. Ltd.

FOR M. F. S.

PREFACE

This study guide was written to accompany PHYSICS: PRINCIPLES WITH APPLICATIONS, Sixth Edition, Vol. 1 by Douglas C. Giancoli. It is intended to provide additional help in understanding the basic principles covered in the textbook and add to the student's problem solving skills.

Each chapter begins with a list of Course Objectives based on the information covered in the chapter. This section is followed by a list of Key Terms and Phrases and a list of the basic mathematical equations used in the textbook. The Concept Summary section of the study guide summarizes the main topics covered in the corresponding chapter of the textbook. The concept summary section also includes the answers to three or four End-Of-Chapter questions from the textbook as well as four or five example problems similar to the type of problems found in the textbook. Each chapter of the study guide concludes with the step-by-step process to the solution to six representative problems taken from the textbook.

Because beginning physics courses emphasize problem solving, hints on problem solving skills are placed just before the representative problems taken from the textbook. The problems are solved using a programmed problem approach. A suggestion for the proper use of the programmed problem method to solving problems is included in chapter 1. The section of the textbook to which each problem corresponds is included as part of the solution.

The student should be aware that the study guide is meant to complement the textbook, not to replace the textbook as a learning tool. Because of this, it is suggested that the student carefully read the chapter in the textbook before using the study guide.

I wish to acknowledge the cooperation and assistance given by Christian Botting, Associate Editor, Physics and Astronomy, Prentice Hall, Inc. Every effort has been made to avoid errors; however, I alone have responsibility for any errors which remain and corrections and comments are most welcome.

Joseph J. Boyle
Professor Emeritus
Miami-Dade College

CONTENTS

CHAPTER 1

INTRODUCTION, MEASUREMENT, ESTIMATING

OBJECTIVES

After studying the material of this chapter, the student should be able to:

- distinguish between a scientific model and a scientific theory.
- explain why experiments are important in the testing of a theory and the improvement of a model.
- explain why uncertainty is present in all measurements and state the uncertainty after taking a measurement.
- calculate the percent uncertainty in a measurement.
- state the Système International (SI) units of mass, length, and time.
- state the metric (SI) prefixes multipliers) and use these prefixes in problem solving.
- convert English units to SI units and vice versa and use the factor-label method in problem solving.
- distinguish between basic quantities and derived quantities as well as basic units and derived units.
- express a number in power of ten notation and use power of ten notation in problem solving.
- explain what is meant by an order-of-magnitude estimate and use order-of-magnitude estimates in problems involving rapid estimating.

KEY TERMS AND PHRASES

science attempts to explain natural phenomena that can be detected with our senses or by instruments designed to extend our senses, e.g., a telescope.

physics is the branch of science that deals with natural laws and processes, and the states and properties of matter and energy.

matter refers to any object that has substance and occupies space.

energy is a measure of the ability to do work. Energy takes a number of forms, such as mechanical energy, electromagnetic energy, heat energy, and nuclear energy.

model, as used in physics, is an analogy or mental image used to explain physical phenomena.

scientific theory, as used in physics, is a plausible principle offered to explain physical phenomenon. A theory leads to predictions that can then be tested by experiment to see if there is agreement with the phenomenon.

scientific law is applied to certain statements that are found to be valid over a large range of observed phenomena. An example of a scientific law is the law of universal gravitation.

significant figures in a measurement include the figures that are certain plus the first doubtful digit.

Système International (SI) system of measurement is the system of measurement established by the French Academy of Science. For example, in SI units, the unit of length is the meter (m), time is second (s), and mass is kilogram (kg).

order-of-magnitude is a rough estimate of the value of a quantity. This estimate usually contains one significant figure and the associated power of ten. For example, one often hears the world population given to the nearest billion.

CONCEPT SUMMARY

Science and Creativity

Science attempts to explain natural phenomena that can be detected with our senses or by instruments designed to extend our senses, e.g., a telescope. Science is a creative endeavor that resembles other creative activities of the human mind, e.g., art and music. **Physics** is the branch of science that deals with natural laws and processes, and the states and properties of **matter** and **energy**.

To explain a particular natural phenomenon, a scientist constructs a **model** that leads to a **theory** designed to explain the phenomenon. The theory leads to predictions that can then be tested by experiment to see if there is agreement with the phenomenon.

In science, the term **law** is applied to certain statements that are found to be valid over a large range of observed phenomena. An example of a scientific law is the law of universal gravitation. Scientific laws are descriptive in that they "describe how nature does behave" as compared to a traffic law which tells us how we should behave.

Measurement and Uncertainty

Every measurement is limited in terms of accuracy. This limitation is associated with the measuring instrument and as stated in the textbook human "inability to read the instrument beyond some fraction of the smallest division shown." Because of this, it is common to include the estimated uncertainty associated with a scientific measurement. For example, the width of a table might be 85.10 ± 0.01 inches. The ±0.01 inches is the uncertainty in the measurement. The percent uncertainty is the ratio of the uncertainty to the measured value, for example, 0.01/85.10 x 100% = 0.01%.

The number of significant figures in a measurement includes the figures that are certain and the first doubtful digit. In the example of the table, which is 85.10 ± 0.01 inches long, there are four significant figures. The 8, 5, and 1 are certain and the 0 is the doubtful digit. Calculations must follow the rules for significant figures. The final answer must have the same number of significant figures as the least significant factor used in the calculation.

TEXTBOOK QUESTION 4. What is wrong with this road sign: Memphis 7 mi (11.263 km)?

ANSWER: The road sign states the distance in miles to one significant figure while it states the distance in kilometers to 5 significant figures. The same number of significant figures should be used whether the distance is given in miles or kilometers.

HOW TO USE THE PROGRAMMED PROBLEM APPROACH TO PROBLEM SOLVING

The example problems and problems from the textbook follow a programmed problem method of solving problems. The problems are arranged in a step-by-step process that leads to the solution. The student should be aware that there is often more than one way to solve a problem and at times an alternate solution will be included.

Each part of the problem is broken down into individual steps. The steps are represented by individual frames that are divided into a left side where a question is posed and a right side where the answer is located. Each step is designated with a letter and number. For example, Part b. Step 2. indicates that this frame is the second step to the solution of Part b of the problem. The right side should be covered with a blank sheet of paper on which you should attempt to answer the question. After completing your answer, uncover the right frame and check your work. The individual frames are separated by a line that extends across the page and the end of the problem is indicated by a thick black line.

EXAMPLE PROBLEM 1. Determine the percent uncertainty in the area of a square that is 6.08 ± 0.01 m on a side.

Part a. Step 1. Determine the area if each side is 2.54 m in length.	Solution: (Section 1-4) area = 6.08 m x 6.08 m = 37.0 m^2 Note: using a calculator, the product is 36.9664. However, there are only three significant figures in each factor and the answer must be rounded off to three significant figures.
Part a. Step 2. Determine the maximum possible area.	maximum length = 6.08 m + 0.01 m = 6.09 m maximum area = 6.09 m x 6.09 m = 37.1 m^2
Part a. Step 3. Determine the minimum possible area.	minimum length = 6.08 m - 0.01 m = 6.07 m minimum area = 6.07 m x 6.07 m = 36.9 m^2
Part a. Step 4. Write the area with the associated uncertainty.	The maximum area is greater than the area by 0.10 m^2 and the minimum area is less by 0.10 m^2. The area can now be written as area = 37.0 ± 0.10 m^2

Part a. Step 5. Determine the percent uncertainty.	percent uncertainty = $(0.10\ m^2)/(37.0\ m^2) \times 100\%$ Percent uncertainty = 0.27%

Système International (SI) System of Measurement

The measurement of any quantity is made relative to a particular standard or unit. The system used almost exclusively in this book is the Système International (SI system of measurement). In SI units, the unit of length is the meter (m), time is second (s), and mass is kilogram (kg). Length, time, mass, electric current, temperature, amount of substance, and luminous intensity are base quantities. The base unit associated with electric current is the ampere, temperature is degrees kelvin, amount of substance is the mole, and luminous intensity is candela. All other quantities can be derived from the base quantities. The units associated with these derived quantities are called derived units. An example of a derived quantity is force and the unit of force is the newton (N) where $1\ N = 1\ kg\ m/s^2$.

TEXTBOOK QUESTION 1. What are the merits and drawbacks of using a person's foot as a standard? Consider both (a) a particular person's foot and (b) any person's foot. Keep in mind that it is advantageous that fundamental standards be accessible (easy to compare to), invariable (do not change), indestructible, and reproducible.

ANSWER: For a standard to be useful it must be accessible, invariable, indestructible, and reproducible everywhere. The problem that is now faced is that while the criteria invariable and indestructible might be met, the fundamental standards may not be accessible or easily understood by the average person.

For example, from 1893 to 1960 the United States standard for length was the distance between two scratches on a metal bar made of a platinum-iridium alloy. This standard meets the criteria plus has the advantage of being easily understandable to the common person. Since 1960, the meter has been defined as 1,650,763.73 times the wavelength of the red-orange spectral line of the krypton-86 isotope. This standard meets the criteria but is less accessible and certainly less understandable to the average person.

Whether the standard is the foot of a particular person or any person's foot, the major drawback is that the length of the foot tends to change with time and is not reproducible. Also, the standard is lost and therefore destructible when the person dies. The only advantage is that a foot is accessible and can be easily used to give a rough estimate of short distances.

EXAMPLE PROBLEM 2. Express the height of a person 5′4″ tall in a) centimeters and b) meters. Express your answer to the correct number of significant figures.

Part a. Step 1. Convert 5′4″ to inches.	Solution: (Section 1-6) (5 feet)(12 inches/1 feet) + 4 inches = 64 inches

Part a. Step 2.	(64 inches)(2.54 cm/1 inch) = 162.6 cm
Express the person's height in centimeters.	64 inches has only two significant figures; therefore, 162.6 cm must be reduced to two significant figures. The person's height to the correct number of significant figures is 160 cm.
Part b. Step 1.	(160 cm)(1.00 m/100 cm) = 1.60 m
Express the height in meters (m).	The answer must be reduced to two significant figures. The person's height is 1.6 m.

EXAMPLE PROBLEM 3. The equatorial diameter of the Earth given to three significant figures is 7930 miles. Express the diameter in a) meters (m) and b) kilometers (km) and use power of ten notation in your final answer.

Part a. Step 1.	Solution: (Section 1-6)
Express the diameter in meters (m). Note: 1 mile = 1609 m	Using a calculator, the diameter is (7930 mi)(1609 m/1 mi) = 12,759,370 m The diameter of the Earth is given to three significant figures. It is necessary to round off the answer to three significant figures. Therefore, the diameter = 12,800,000 m The diameter in power of ten notation = 1.28×10^7 m
Part b. Step 1.	1.000 km = 1000 m
Determine the diameter in km.	(1.28×10^7 m)(1.000 km/1000 m) = 1.28×10^4 km

Order of Magnitude: Rapid Estimating

It is sometimes useful to give a rough estimate of the value of a quantity. This estimate is called the **order-of-magnitude** and this number contains one significant figure and the associated power of ten. For example, one often hears the world population given to the nearest billion.

An example of rapid estimating would be the number of people watching a parade. Suppose the parade route is 1 mile long with people lined up 5 deep on each side of the street. Each person taking up an average of 2 feet of space.

(1 person/2 feet) x (5280 feet/1 mile) x 2 sides of street x (5 people/1 space) $\approx$ 30,000 people

The symbol $\approx$ means that the measurement is an approximation. For this problem, the actual size of the crowd could easily be $\pm$3000 to 5000 people.

PROBLEM SOLVING SKILLS

For problems involving percent uncertainty in area or volume and the linear dimensions are given:

1. Determine the maximum and minimum possible values for the area or volume.
2. Determine the uncertainty in the measurement.
3. Divide the uncertainty by the measurement and multiply by 100%. If the problem involves a linear measurement, then steps 1 and 2 are not necessary to solve the problem.

For problems involving conversion of English units to SI units or vice versa.

1. List the quantity given and, if necessary, express it in one unit. For example, 1 min and 10 seconds = 70 seconds.
2. Use the unit rule, also known as the factor-label method, to solve the problem.
3. Where appropriate, use power of ten notation. Make sure to follow the rules for multiplication, division, addition, and subtraction of quantities which are expressed as powers of ten.
4. Apply the rules of significant figures in solving the problem. Remember that the answer must have the same number of significant figures as the least significant factor.

SOLUTIONS TO SELECTED TEXTBOOK PROBLEMS

TEXTBOOK PROBLEM 1. The age of the universe is thought to be about 14 billion years. Assuming two significant figures, write this in powers of ten in (a) years, (b) seconds.

Part a. Step 1.	Solution: (Section 1-4)
Use the factor-label method to convert to power of ten.	1 billion = 10^9 (14 billion years)[(10^9 years)/(1 billion)] = 1.4×10^{10} years
Part a. Step 2.	(1.4×10^{10} yr)[(365.25 days)/(1 yr)][(24 h)/(1 day)][(3600 s)/(1 hour)]
Use the factor-label method to convert years to seconds (s).	= 4.4×10^{17} s

TEXTBOOK PROBLEM 6. What is the percent uncertainty in the measurement 3.76 ± 0.25 m?

Part a. Step 1.	Solution: (Section 1-4)
Determine the percent uncertainty.	The percent uncertainty is the ratio of the uncertainty to the measured value multiplied by 100 %. Therefore, (0.25 m)/(3.76 m) x 100% = 6.6 %

TEXTBOOK PROBLEM 15. The Sun, on average, is 93 million miles from the Earth. How many meters is this? Express (a) using powers of ten, and (b) using a metric prefix.

Part a. Step 1. Use the factor-label method to convert miles to meters. 1609 m = 1 mile	Solution: (Section 1-6) (93 million miles)[(10^6 miles)/(1 million miles)][(1609 m)/(1 mile)] = 1.5×10^{11} m
Part b. Step 1. Express the answer using a metric prefix.	From Table 1-4 the prefix giga, abbreviated G, represents 10^9. Therefore, (1.5×10^{11} m)[(1 G)/(10^9)] = 1.5×10^2 Gm or 150 Gm

TEXTBOOK PROBLEM 23. The diameter of the Moon is 3480 km. What is the surface area of the Moon. (b) How many times larger is the surface area of the Earth?

Part a. Step 1. Determine the is surface area of the Moon.	Solution: (Sections 1-5 and 1-6) Assume that the Moon is a sphere. The formula for the surface area (A) of a sphere is $A = 4\pi r^2$ where r = the radius of the sphere. r = ½ diameter = ½ (3480 km) = 1740 km $A_{Moon} = 4\pi (1740 \text{ km})^2 = 3.80 \times 10^7 \text{ km}^2$
Part b. Step 1. Determine the land surface area of the Earth.	Based on information provided in the inside cover of the textbook, the radius of the Earth is 6380 km. $A_{Earth} = 4\pi (6380 \text{ km})^2 = 5.11 \times 10^8 \text{ km}^2$
Part b. Step 2. Determine the ratio the surface areas.	ratio of areas = $(5.11 \times 10^8 \text{ km}^2)/(3.80 \times 10^7 \text{ km}^2) \approx 13.4$ The Earth has approximately 13.4 times more surface area than the of Moon.

TEXTBOOK PROBLEM 27. Estimate how long it would take one person to mow a football field using an ordinary home lawn mower (Fig. 1-13). Assume that the mower moves with a 1 km/h speed, and has a 0.5 m width.

Part a. Step 1. Determine the lawn mower's speed in m/s.	Solution: (Section 1-7) (1 km/h) x (1000 m/km) x (1h/3600 s) $\approx$ 0.28 m/s

Part a. Step 2. Estimate the area cut by the mower each second.	The area cut each second equals the forward speed times the length of the blade. $(0.28\ m/s)(0.5\ m) \approx 0.14\ m^2/s$
Part a. Step 3. Estimate the area of a football field.	Length of the football field = 100 yards + 10 yards per end zone Length = 120 yards ≈ 110 meters Width of the football field is 53.3 yards ≈ 49 m Area = length x width = 110 m x 49 m ≈ 5400 m^2
Part a. Step 4. Estimate the time required to mow the field.	time = (area of football field) ÷ (area cut each second) time = $(5400\ m^2) \div (0.14\ m^2/s) \approx 39000\ s \approx 11$ hours Note: this is a very rough estimate. The estimate assumes that the person continues to mow at a constant rate for the entire time.

TEXTBOOK PROBLEM 51. The diameter of the Moon is 3480 km. What is the volume of the Moon? How many Moons would be needed to create a volume equal to the volume of the Earth?

Part a. Step 1. Determine the volume of the Moon. Assume that the Moon is a perfect sphere.	Solution: (Section 1-7) $V_{Moon} = 4/3\ \pi\ R^3$ where $R_{Moon} = ½(3480\ km) = 1740\ km$ $= 4/3\ \pi\ [(1740\ km)(1000\ m/1\ km)]^3$ $V_{Moon} = 2.20 \times 10^{19}\ m^3$
Part a. Step 2. Determine the ratio of the Earth's volume to the Moon's volume. Note: the radius of the Earth is 6380 km.	$V_{Earth}/V_{Moon} = (4/3\ \pi\ R_E^3)/(4/3\ \pi\ R_M^3)$ Note: both 4/3 and π cancel. $= R_E^3/R_M^3$ $= (6380\ km)^3/(1740\ km)^3$ $V_{Earth}/V_{Moon} \approx 49.3$

CHAPTER 2

DESCRIBING MOTION: KINEMATICS IN ONE DIMENSION

OBJECTIVES

After studying the material of this chapter, the student should be able to:

- state from memory the meaning of the key terms and phrases used in kinematics.
- list the SI unit and its abbreviation associated with displacement, velocity, acceleration, and time.
- describe the motion of an object relative to a particular frame of reference.
- differentiate between a vector quantity and a scalar quantity and state which quantities used in kinematics are vector quantities and which are scalar quantities.
- state from memory the meaning of the symbols used in kinematics: x, x_o, v, v_o, a, y, y_o, v_y, v_{yo}, g, t.
- write from memory the equations used to describe uniformly accelerated motion.
- complete a data table using information both given and implied in word problems.
- use the completed data table to solve word problems.
- use the methods of graphical analysis to determine the instantaneous acceleration at a point in time and the distance traveled in an interval of time.

KEY TERMS AND PHRASES

kinematics is the study of the motion of objects and involves the study of distance, speed, acceleration, and time.

average speed of an object is determined by dividing the distance that the object travels by the time required to travel that distance.

instantaneous speed is the speed of an object at a particular point in time.

acceleration is the rate of change of speed in time.

average acceleration is the change of velocity divided by the time required for the change.

instantaneous acceleration is the change of speed that occurs in a very small interval of time.

uniformly accelerated motion occurs when the rate of acceleration does not change, i.e., the rate of acceleration is constant.

free fall occurs when air resistance on an object is negligible and the only force acting on it is gravity.

gravitational acceleration for all objects in free fall is approximately 9.8 meters per second per second or 9.8 m/s^2.

vector is a quantity that has both magnitude and direction. Examples of vector quantities are velocity, acceleration, displacement, and force.

scalar is a quantity that has magnitude but has no direction associated with it. Examples of scalar quantities include speed, distance, mass, and time.

SUMMARY OF MATHEMATICAL FORMULAS

average acceleration	$\bar{a} = \Delta v / \Delta t$	The average acceleration equals the change in speed divided by the change in time.
kinematics equations for uniformly accelerated motion	$v = v_o + a t$	Speed as related to initial speed, acceleration and time.
	$x = x_o + v_o t + \frac{1}{2} a t^2$	Distance as related to initial distance, initial speed, acceleration, and time.
	$v^2 = v_o^2 + 2 a (x - x_o)$	Speed as related to initial speed, acceleration, and distance.
	$x - x_o = \bar{v} t$	Distance traveled equals the product of the average speed and the time.
	$\bar{v} = \frac{1}{2}(v + v_o)$	Average speed as related to the initial speed and final speed.

CONCEPT SUMMARY

Kinematics

Kinematics is the study of the motion of objects and involves the study of the following concepts: distance, speed, acceleration, and time. This chapter is restricted to the motion of an object along a straight line. This is known as one-dimensional or linear motion.

The **average speed** of an object is determined by dividing the distance that the object travels by the time required to travel that distance. The **instantaneous speed** refers to the speed of an object at a particular point in time. In this study guide the SI system of units will be used. The SI unit of speed is meters per second (m/s) or kilometers per hour (km/h). SI is an abbreviation of the French words Système International. This system was formerly referred to as the MKS (meter-kilogram-second) system.

Acceleration refers to the rate of change of speed in time. The SI unit of acceleration is meters per second per second or m/s^2. The **average acceleration** is defined as the change of velocity divided by the time required for the change. The **instantaneous acceleration** refers to the change of speed that occurs in a very small interval of time ($\Delta t \rightarrow 0$).

Uniformly Accelerated Motion

Uniformly accelerated motion occurs when the rate of acceleration does not change, i.e. the rate of acceleration is constant. The following equations apply to this type of motion:

$v = v_o + a\,t$ $\qquad$ $x = x_o + v_o\,t + \tfrac{1}{2}\,a\,t^2$

$v^2 = v_o^2 + 2\,a\,(x - x_o)$ $\qquad$ $x - x_o = \bar{v}\,t$ where $\bar{v} = \tfrac{1}{2}(v + v_o)$

v_o = initial speed of the object (at t = 0 s) $\qquad$ v = speed of the object after time t

a = rate of acceleration $\qquad$ x = position of the object after time t

$\bar{v}$ = average speed $\qquad$ x_o = initial position of the object
Note: $x_o = 0$ unless otherwise specified.

t = time interval during which the motion has occurred; unless otherwise specified, the time at the start of the motion will be zero seconds.

EXAMPLE PROBLEM 1. A car accelerates from 10.0 m/s to a speed of 30.0 m/s in 10.0 s. If the rate of acceleration is uniform, determine the a) rate of acceleration and b) distance traveled during the 10.0 s of motion.

Part a. Step 1. List each symbol and complete a data table based on the information given in the problem.	Solution: (Sections 2-5 and 2-6) $v_o = 10.0$ m/s $\quad$ $t = 10.0$ s $\quad$ $v = 30.0$ m/s $a = ?$ $\quad$ $x = ?$ $\quad$ $x_o = 0$
Part a. Step 2. Determine the rate of acceleration.	$v = v_o + a\,t$ $30 \text{ m/s} = 10.0 \text{ m/s} + a\,(10.0 \text{ s})$ $a = 2.0 \text{ m/s}^2$
Part b. Step 1. Determine the distance traveled in 10.0 seconds.	$x = x_o + v_o\,t + \tfrac{1}{2}\,a\,t^2$ $x = 0 \text{ m} + (10.0 \text{ m/s})(10.0 \text{ s}) + \tfrac{1}{2}(2.0 \text{ m/s}^2)(10.0 \text{ s})^2$ $x = 200$ m

EXAMPLE PROBLEM 2. A car is traveling at the posted speed limit of 15.0 miles per hour (6.70 m/s) in a school zone on dry pavement. The driver applies maximum braking force and the car decelerates at a constant rate of 7.92 m/s^2 until coming to a halt. a) Calculate the distance the car travels while decelerating. b) Calculate the distance required to stop if the car is initially traveling at 30.0 miles per hour (13.4 m/s).

Part a. Step 1. Complete a data table for the deceleration.	Solution: (Sections 2-5 and 2-6) $x = ?$ $x_o = 0$ m $a = -7.92$ m/s^2 $v_o = 6.70$ m/s $v = 0$ m/s (car comes to a halt) $t = ?$
Part a. Step 2. Calculate the distance the car would travel during the deceleration.	$2a(x - x_o) = v^2 - v_o^2$ $2(-7.92 \text{ m/s}^2)(x - x_o) = (0 \text{ m/s})^2 - (6.70 \text{ m/s})^2$ $(-15.8 \text{ m/s}^2)(x - x_o) = -44.9 \text{ m}^2/\text{s}^2$ $x - x_o = 2.83$ m or 9.30 feet
Part b. Step 1. Complete a data table for the deceleration.	$x = ?$ $x_o = 0$ m $a = -7.92$ m/s^2 $v_o = 13.4$ m/s $v = 0$ m/s (car comes to a halt) $t = ?$
Part b. Step 2. Calculate the distance the car would travel during the deceleration.	$2a(x - x_o) = v^2 - v_o^2$ $2(-7.92 \text{ m/s}^2)(x - x_o) = (0 \text{ m/s})^2 - (13.4 \text{ m/s})^2$ $(-15.8 \text{ m/s}^2)(x - x_o) = -180 \text{ m}^2/\text{s}^2$ $x - x_o = 11.3$ m or 37.4 feet At 30.0 mph, the car requires approximately four times more distance to come to a complete stop as compared to 15.0 mph. Therefore, in order to protect children, a low speed limit is set when traveling through a school zone. Using the same method it can be shown that at 60 miles per hour the stopping distance is 16 times further than at 15 mph, i.e., 149 feet. On wet pavement the stopping distance is even greater.

EXAMPLE PROBLEM 3. A car traveling at a constant speed of 15.0 m/s (approximately 34 miles per hour) in a zone where the posted speed limit is 25.0 miles per hour. As the motorist passes a stationary police car, the police car accelerates at a constant rate of 3.00 m/s^2 and maintains this rate of acceleration until the police car pulls next to the speeding car. Determine the (a) time required for the police officer to catch the speeder and (b) distance traveled during the chase. Hint: In order to catch the speeder the police car must exceed 15.0 m/s.

Part a. Step 1. Complete a data table for both vehicles using information both given and implied in the statement of the problem.	Solution: (Sections 2-5 and 2-6) motorist's car — police car $v_o = 15.0$ m/s — $v = 0$ m/s $v = 15.0$ m/s — $v = ?$ $a = 0$ m/s — $a = 3.00$ m/s^2

	$t = ?$	$t = ?$
	$x = ?$	$x = ?$
	$x_o = 0$ m	$x_o = 0$ m

Part a. Step 2. Write an equation for the motorist's car as a function of time.	motorist $x - x_o = \overline{v}\, t$ but $x_o = 0$, therefore $x = \frac{1}{2}(15.0 \text{ m/s} + 15.0 \text{ m/s})t$ and $x = (15.0 \text{ m/s})t$
Part a. Step 3. Write an equation for the position of the police car as a function of time.	police car $x = v_o t + \frac{1}{2} a t^2 + x_o$ $x = (0 \text{ m/s})\, t + \frac{1}{2} (3.00 \text{ m/s}^2)\, t^2 + 0 \text{ m}$ $x = (1.50 \text{ m/s}^2)\, t^2$
Part a. Step 4. Determine the time required for the police to catch the motorist. Note: when the police car pulls alongside the motorist, it is traveling faster than the motorist's car. However, both cars have traveled the same distance.	Since $x_{motorist} = x_{police\ car}$ $(15.0 \text{ m/s})\, t = (1.50 \text{ m/s}^2)\, t^2$ writing the expression as an algebraic equation and solving for t, $15.0\, t = 1.50\, t^2$ $0 = 1.5\, t^2 - 15.0\, t$ $0 = (t - 10.0)(1.50\, t)$ Either $0 = t - 10.0$ or $0 = 1.50\, t$ $t = 10.0$ s or $t = 0$ s The equation is a quadratic equation and two values are obtained for the time of motion. However, $t = 0$ s merely corresponds with our initial assumption that the two cars were at the same position at $t = 0$. The correct solution is $t = 10.0$ s.
Part b. Step 1. Determine the distance each vehicle travels during the time interval.	Add $t = 10.0$ s to the data table and solve for the distance traveled during the chase. $x - x_o = \overline{v}\, t$ $x - 0 \text{ m} = (15.0 \text{ m/s})(10.0 \text{ s})$ $x = 150$ m

Free Fall

One application of uniformly accelerated motion is to the problem of objects in **free fall.** When an object is in free fall, we assume that air resistance is negligible and that the only force acting on it is gravity. Assuming air resistance is negligible, the rate of acceleration (g) of all objects in free fall is approximately 9.8 meters per second per second or 9.8 m/s^2.

The equations for uniformly accelerated motion can be applied to free fall. Since the motion is vertical, y replaces x and y_o replaces x_o. Also, v_y replaces v and v_{yo} replaces v_o while g replaces the symbol a.

TEXTBOOK QUESTION 14. How would you estimate the maximum height you could throw a ball vertically upward? How would you estimate the maximum speed you could give it?

ANSWER: Suppose the total time for the ball to return to your hand is measured to be 4.0 s. Then the time to reach maximum height is ½(4.0 s) = 2.0 s. Also, as the ball rises it is decelerating, i.e. g = -9.8 m/s^2 and at maximum height its speed is zero, i.e., $v_y = 0$. The ball's initial speed can be calculated as follows:

$$v_y = g\,t + v_{oy}$$

$$0 = (-9.8\ m/s^2)(2.0\ s) + v_{oy} \quad \text{and} \quad v_{oy} = 19.6\ m/s$$

The maximum height reached by the ball can be determined as follows:

$$y = v_o\,t + \tfrac{1}{2}\,a\,t^2$$

$$y = (19.6\ m/s)(2.0\ s) + \tfrac{1}{2}\,(-9.8\ m/s^2)(2.0\ s)^2$$

$$y = 19.6\ m$$

EXAMPLE PROBLEM 4. A stone is thrown vertically downward with an initial speed of 9.80 m/s from the top of a building 29.4 m high. Determine the a) velocity of the stone just before it strikes the ground and b) time that the stone is in the air.

Part a. Step 1.	Solution: (Section 2-7)	
Because the motion is downward, let the downward direction be positive. Complete a data table based on the information given.	$v_o = 9.80\ m/s$	$t = ?$
	$v = ?$	$y = 29.4\ m$
	$g = 9.80\ m/s^2$	$y_o = 0\ m$
Part a. Step 2.	$v^2 = v_o^2 + 2\,g\,(y - y_o)$	
Determine the velocity of the stone just before it strikes the ground.	$v^2 = (9.80\ m/s)^2 + 2(9.80\ m/s^2)(29.4\ m - 0\ m)$	
	$v^2 = 96.0\ m^2/s^2 + 576\ m^2/s^2$	

Note: after it leaves the person's hand it is in free fall.	$v^2 = 672\ m^2/s^2$ $v = 25.9\ m/s$
Part b. Step 1. Determine the time that the stone is in the air.	The velocity of the stone just before it strikes the ground can now be included in the data table. $v = v_o + a\,t$ $25.9\ m/s = 9.80\ m/s + (9.80\ m/s^2)\ t$ $t = 1.64\ s$

Frames of Reference

The description of motion of any object must always be given relative to a **frame of reference** or **reference frame**. The reference frame is usually specified by using **Cartesian Coordinates**. The x, y, and z axes shown in the figures below can be used to locate the position of an object with respect to a fixed point o. At times, the x and y directions will be used to represent the direction of cardinal points: north (N), south (S), east (E), and west (W), with "up" above the plane of the paper and "down" below the plane of the paper. In certain problems, it will be convenient to use the x axis to represent the horizontal direction while the y axis represents the vertical direction.

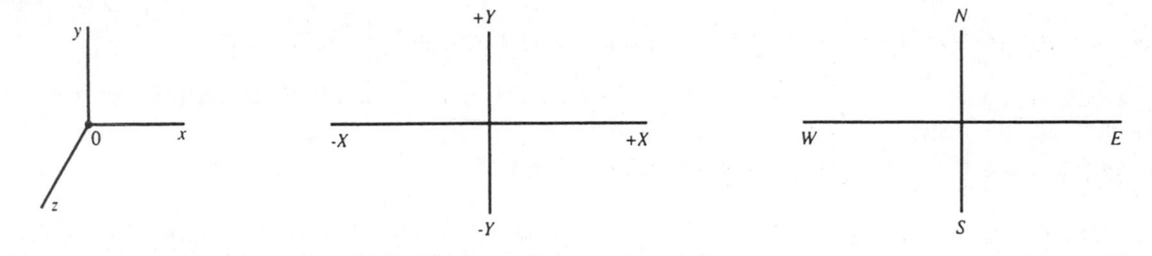

Vectors and Scalars

A **vector** is a quantity that has both magnitude and direction. An example of a vector quantity is velocity. A car traveling at 20 m/s must be traveling in a specified direction, e.g., due north. Displacement, velocity, and acceleration are examples of vector quantities. A vector is designated by the symbol for the vector quantity in bold face type as in the textbook. For example, the velocity vector will be represented by a bold face **v** with an arrow placed above it, i.e., $\vec{\mathbf{v}}$.

A **scalar** is a quantity that has magnitude but has no direction associated with it. A scalar is specified by giving its magnitude and units (if any). Speed, distance, mass, and time are examples of scalar quantities. As an example, speed refers to the magnitude of an object's motion but not the direction in which it is traveling.

TEXTBOOK QUESTION 1. Does a car speedometer measure speed, velocity, or both?

ANSWER: A car's speedometer indicates the magnitude of the car's motion. It does not indicate the car's direction. Since velocity is both magnitude and direction, the speedometer measures speed but does NOT measure velocity.

TEXTBOOK QUESTION 2. Can an object have varying speed if its velocity is constant? If yes, give examples.

ANSWER: Velocity is a vector quantity while speed is a scalar quantity. In order for the velocity to be constant both the magnitude, i. e. speed, as well as the direction must be constant. Therefore, if the speed changes then the velocity is not constant.

TEXTBOOK QUESTION 7. Can an object have a northward velocity and a southward acceleration? Explain.

ANSWER: If the object is traveling northward, then the direction of the velocity vector is northward. If the object slows, but is still traveling northward, then the velocity vector is northward but the car is decelerating. Acceleration (or deceleration) is a vector quantity. If the object is accelerating, then the acceleration vector is in the same direction as the velocity vector. In the instance where the object is decelerating, the acceleration vector is directed opposite from the velocity vector. Therefore, for an object traveling northward but decelerating, the velocity vector is northward but the acceleration vector is southward.

Graphical Analysis

Graphs can be used to analyze the straight line motion of objects. Although **graphical analysis** can be used for uniformly accelerated motion, the method is especially useful when dealing with the motion of an object that is not undergoing uniform acceleration.

In a distance versus time graph the instantaneous velocity at any point can be determined from the slope of a tangent line drawn to the point in question. In a velocity versus time graph the slope of the tangent line represents the instantaneous acceleration while the area under the curve represents the distance traveled.

EXAMPLE PROBLEM 5. The graph shown below represents the motion of a car over a period of 10.0 s. Use graphical analysis to determine the a) rate of acceleration of the car at 2.0 s, 5.0 s, and 8.0 s and b) distance traveled during the 10.0 s of motion.

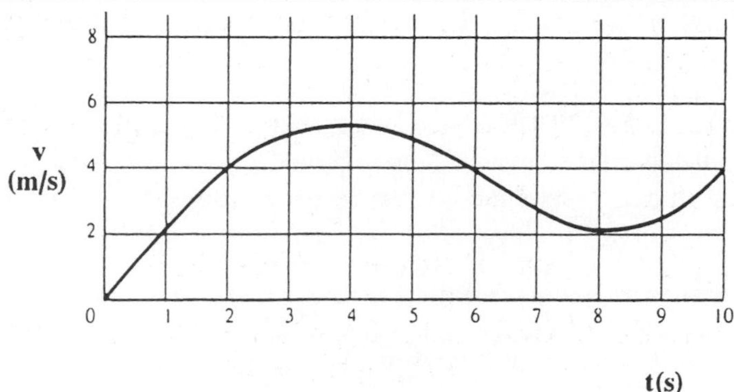

Part a. Step 1.

Determine the acceleration of the car at 2.0 s, 5.0 s, and 8.0 s.

Solution: (Section 2-8)

To determine the rate of acceleration, a tangent line is drawn at each point in question. Two data points are selected from each tangent line and the rate of acceleration is determined by using the following formula:

$$a = (v_2 - v_1)/(t_2 - t_1)$$

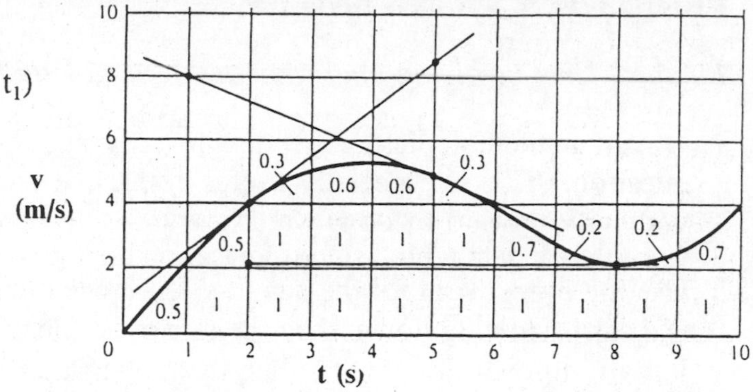

at t = 2.0 s: $a = (8.5 \text{ m/s} - 4.0 \text{ m/s})/(5.0 \text{ s} - 2.0 \text{ s}) = +1.5 \text{ m/s}^2$

at t = 5.0 s: $a = (8.0 \text{ m/s} - 5.0 \text{ m/s})/(1.0 \text{ s} - 5.0 \text{ s}) = -0.75 \text{ m/s}^2$

at t = 8.0 s: $a = (2.2 \text{ m/s} - 2.2 \text{ m/s})/(8.0 \text{ s} - 2.0 \text{ s}) = 0 \text{ m/s}^2$

There is a certain amount of judgment required in drawing the tangent lines. The actual value of the acceleration at each point may be different from the values obtained above. Your knowledge of algebra will help in checking whether or not the answers are reasonable. The tangent line at t = 2.0 s has a positive slope and therefore the car should be accelerating. The car's speed is increasing at this point; therefore, it is accelerating. At t = 5.0 s, the car is slowing down and is therefore decelerating. The slope of the tangent line at t = 5.0 s is negative and therefore agrees with observation. The slope of the tangent line at t = 8.0 s is zero. This indicates that the car is neither accelerating nor decelerating. Looking at the graph, we can see that the car has stopped slowing down and has not yet begun to increase in speed; thus its rate of acceleration is indeed zero.

Part b. Step 1.

Use graphical integration to determine the distance traveled during the 10.0 s motion.

The distance traveled can be determined by calculating the area under the graph. This can be done by determining the distance represented by the area of one block and then multiplying this value by the total number of full and partial blocks that lie between the curve and the time axis. Note: a major source of error in determining the distance traveled is the judgment required in estimating the value of a partial block. A more accurate value can be obtained by using graph paper that contains a fine grid.

2.0 m/s × 1.0 s block

distance represented by one block = (2.0 m/s)(1.0 s)

distance = 2.0 m

sum of complete blocks = 13.0

sum of partial blocks = 0.5 + 0.5 + 0.3 + 0.6 + 0.6 + 0.3 + 0.7 + 0.2 + 0.2 + 0.7 = 4.6

total number of blocks = 13.0 + 4.6 = 17.6

total distance = (17.6 blocks)(2.0 m/1 block) = 35 m

PROBLEM SOLVING SKILLS

For problems involving uniformly accelerated motion:

1. Obtain a mental picture by drawing a diagram that reflects the motion of the object in question. This is especially useful in the free fall problems where the initial motion may be vertically upward or downward.
2. Complete a data table using information both given and implied in the wording of the problem.
3. Use the proper sign for the quantity represented by the symbol in the data table. For example, if a car is slowing down, then the rate of acceleration is negative. If an object in free fall was initially thrown downward, then the downward direction is taken to be positive and both the initial velocity and the rate of acceleration are positive. If the object in free fall was given an initial upward motion, then the upward direction is taken to be positive. This means that the initial upward velocity is positive but the rate of acceleration is negative because it is slowing down as it travels upward.
4. Memorize the formulas for uniformly accelerated motion. It is also necessary to memorize the meaning of each symbol in each formula. Using the data from the completed data table, determine which formula or combination of formulas can be used to solve the problem.

For problems related to graphical analysis where velocity is a function of time:

1. Determine the area of one block. This area represents the distance represented by a single block.
2. Count the number of blocks in the time interval being considered. Multiply the total number of blocks by the distance represented by one block. The product is the total distance traveled during the time interval. This technique is known as graphical integration.
3. The instantaneous acceleration at a particular moment of time is determined as follows: a) draw a tangent line to the graph at the point in question, and b) determine the magnitude of the slope of the tangent line. The magnitude of the slope of the line represents the instantaneous value of the acceleration at that moment in time. If the slope is positive, the object is accelerating. If the slope is negative, the object is decelerating. If the slope is zero, the object is traveling at constant speed.

For problems related to graphical analysis where distance is a function of time:

1. The instantaneous velocity at a particular moment in time equals the slope of the tangent line drawn to the curve at the point in question. If the slope is positive, the object's speed is positive. If the slope is negative, the object's speed is negative. If the slope is zero, the object is not moving.

SOLUTIONS TO SELECTED TEXTBOOK PROBLEMS

TEXTBOOK PROBLEM 9. A person jogs 8 complete laps around a quarter-mile track in a total time of 12.5 min. Calculate (a) the average speed and (b) the average velocity in m/s.

Part a. Step 1. Determine the total distance traveled in meters.	Solution: (Sections 2-1 through 2-3) (8)(¼ mile)(1609 m/mile) = 3220 m
Part a. Step 2. Determine the average speed.	average speed = total distance/total time $\overline{v}$ = (3220 m)/[(12.5 min)(60 s/1 min)] = 4.3 m/s
Part b. Step 1. Determine the runner's total displacement from the starting point.	While distance is a scalar quantity, displacement is a vector quantity. Assuming the runner ended up at the exact spot where he/she started, the runner's displacement is zero.
Part b. Step 2. Determine the runner's average velocity.	As stated in Part b. Step 1, the runner's displacement from the starting point is zero. The runner's average velocity equals the displacement/time. Therefore, the average velocity equals zero. $\overline{v} = \Delta x/\Delta t$ = ($\overline{0}$ m)/[(12.5 min)(60 s/ min)] = 0 m/s

TEXTBOOK PROBLEM 18. At highway speeds, a particular automobile is capable of an acceleration of about 1.6 m/s^2. At this rate, how long would it take to accelerate from 80 km/h to 110 km/h?

Part a. Step 1. Convert km/h to m/s.	Solution: (Section 2-5 and 2-6) v_o = (80 km/h)(1000 m/km)(1 h/3600 s) = 22.2 m/s v = (110 km/h)(1000 m/km)(1 h/3600 s) = 30.5 m/s
Part a. Step 2. Complete a data table.	v_o = 22.2 m/s t = ? a = 1.6 m/s^2 v = 30.6 m/s $x - x_o$ = ?
Part a. Step 3. Select the appropriate equation and solve for the time.	$v = v_o + a t$ 30.5 m/s = 22.2 m/s + (1.6 m/s^2) t t = 5.2 s

TEXTBOOK PROBLEM 26. In coming to a stop, a car leaves a skid mark 92 m long on the highway. Assuming a deceleration of 7.00 m/s^2, estimate the speed of the car just before braking.

Part a. Step 1. Complete a data table based on information both given and implied in the problem.	Solution: (Section 2-5 and 2-6) v_o = ? t = ? a = - 7.00 m/s^2 v = 0 m/s (car comes to a halt) $x - x_o$ = 92 m

Part a. Step 2.

Calculate the initial speed.

$v^2 = v_o^2 + 2\ a\ (x - x_o)$

$(0\ m/s^2) = v_o^2 + 2\ (-\ 7.00\ m/s^2)(92\ m)$

$v_o^2 = 1290\ m^2/s^2$

$v_o = 36\ m/s$

TEXTBOOK PROBLEM 32. A person driving her car at 45 km/h approaches an intersection just as the traffic light turns yellow. She knows that the yellow light lasts only 2.0 s before turning red, and she is 28 m from the near side of the intersection (Fig. 2-31). Should she try to stop, or should she speed up to cross the intersection before the light turns red? The intersection is 15 m wide. Her car's maximum deceleration is -5.8 m/s², whereas it can accelerate from 45 km/h to 65 km/h in 6.0 s. Ignore the length of her car and her reaction time.

Part a. Step 1.

Complete a data table if she stops. Convert the initial speed to m/s.

Solution: (Sections 2-5 and 2-6)

$x = ?$ $\quad x_o = 0\ m$ $\quad t = ?$

$a = -\ 5.8\ m/s^2$ $\quad v = 0\ m/s$ (car comes to a halt)

$v_o = (45\ km/h)[(1000\ m)/(1\ km)][(1\ h)/(3600\ s)] = 12.5\ m/s$

Part a. Step 2.

Calculate the distance the car would travel during the deceleration.

$2\ a\ (x - x_o) = v^2 - v_o^2$

$2\ (-\ 5.8\ m/s^2)(x - x_o) = (0\ m/s)^2 - (12.5\ m/s)^2$

$x - x_o = (-156\ m^2/s^2)/(-\ 11.6\ m/s^2) \approx 13\ m$

The start of the intersection is 28 m from the car's initial position. Based on the calculation, the student could easily stop before reaching the intersection.

Part a. Step 3.

Determine the rate of acceleration if she decides to make it across before the light changes.

$v_o = (45\ km/h)[(1000\ m)/(1\ km)][(1\ h)/(3600\ s)] = 12.5\ m/s$

$v = (65\ km/h)[(1000\ m)/(1\ km)][(1\ h)/(3600\ s)] = 18.1\ m/s$

$v = a\ t + v_o$

$18.1\ m/s = a\ (6.0\ s) + 12.5\ m/s$

$a = 0.926\ m/s^2$

Part a. Step 4.

Determine the distance that the accelerating car would travel in 2.0 s.

$x - x_o = v_o\ t + \frac{1}{2}\ a\ t^2$

$x - x_o = (12.5\ m/s)(2.0\ s) + \frac{1}{2}\ (+0.926\ m/s^2)(2.0\ s)^2$

$x - x_o = 25\ m + 1.85\ m \approx 27\ m$

The intersection is 28 m from her car at the start of the acceleration.

After 2.0 seconds, the car would be just about to enter the intersection traveling at high speed. This means that she would pass through the intersection against a red light. Therefore, she should adjust the rate of deceleration, stop, and wait for the next green light.

TEXTBOOK PROBLEM 36. A baseball is hit nearly straight up into the air with a speed of 22 m/s. (a) How high does it go? (b) How long is it in the air?

Part a. Step 1. Complete a data table based on information both given and implied in the problem.	Solution: (Section 2-7) $v_o = 22$ m/s $t = ?$ $a = -9.80$ m/s^2 $v = 0$ m/s (at top of motion) $y_o = 0$ m $y = ?$ The initial direction of motion is upward, so let the ball's initial speed be positive. The ball decelerates at it rises; therefore, $a = -9.80$ m/s^2. The ball's speed when it reaches maximum height above the ground is zero.
Part a. Step 2. Use the appropriate equation to determine the maximum height reached by the ball.	$2a(y - y_o) = v^2 - v_o^2$ $2(-9.8 \text{ m/s}^2)(y - 0 \text{ m}) = (0 \text{ m/s})^2 - (22 \text{ m/s})^2$ $(-19.6 \text{ m/s}^2)(y - 0 \text{ m}) = -480 \text{ m}^2/\text{s}^2$ $y = 25$ m
Part b. Step 1. Use the appropriate equation to determine the total time the ball is in the air.	The ball started at $y_o = 0$ m and returns to the same position. Therefore, $y = 0$ m. $y - y_o = v_o t + \frac{1}{2} a t^2$ $0 \text{ m} - 0 \text{ m} = (22 \text{ m/s})\, t + \frac{1}{2}(-9.80 \text{ m/s}^2)\, t^2$ $0 \text{ m} = [(22 \text{ m/s}) + (-4.90 \text{ m/s}^2)]\, t$ Solving for the time gives $t = 0$ s or $t = 4.5$ s $t = 0$ s is the time at which it was struck by the bat. Therefore, the time in the air is 4.5 s.

TEXTBOOK PROBLEM 47. A stone is thrown vertically upward with a speed of 12.0 m/s from the edge of a cliff 70.0 m high (Fig. 2-34). (a) How much later does it reach the bottom of the cliff? (b) What is its speed just before hitting? c) What total distance did it travel?

Part a. Step 1. Complete a data table based on information both given and implied in the problem.	Solution: (Section 2-7) v_o = +12.0 m/s t = ? a = - 9.80 m/s^2 v = ? y = -70.0 m The initial direction of motion is upward, so let the stone's initial speed be positive. The stone decelerates at it rises, therefore, a = - 9.80 m/s^2. The stone falls to a point 70.0 m below its starting point; therefore y = - 70.0 m.
Part a. Step 2. Use the appropriate equation to determine the time of flight.	$y = v_o t + \frac{1}{2} a t^2$ $- 70.0\text{ m} = (12.0\text{ m/s})\, t + \frac{1}{2}(-9.80\text{ m/s}^2)\, t^2$ $0 = (-4.90\text{ m/s})\, t^2 + (12.0\text{ m/s})\, t + 70.0\text{ m}$ Using the quadratic formula, it can be shown that either t = 5.20 s or t = -2.74 s Since time cannot be negative, the answer is t = 5.20 s.
Part b. Step 1. The time of flight can now be added to the data table. Solve for the velocity of the stone just before it strikes the ground.	$v = v_o + a t$ $v = 12.0\text{ m/s} + (-9.80\text{ m/s}^2)(5.20\text{ s})$ v = - 38.9 m/s Note: the negative value for the velocity indicates that the stone is traveling downward. As you recall, the upward direction was selected as the positive direction.
Part c. Step 1. Complete a data table based on information both given and implied in the problem. Then determine the height.	v_o = +12.0 m/s t = ? a = - 9.80 m/s^2 v = 0 y = ? The initial direction of motion is upward, so let the stone's initial speed be positive. The stone decelerates at it rises; therefore, a = - 9.80 m/s^2. At maximum height the stone momentarily maximum comes to a halt, i.e., v = 0 m/s. $v^2 = v_o^2 + 2 a (y - y_o)$ $0^2 = (12.0\text{ m/s})^2 + 2(- 9.80\text{ m/s}^2)(y - y_o)$ $y - y_o = 7.35$ m
Part c. Step 2. Determine the total distance traveled.	The stone travels upward 7.35 m. At that point it is 7.35 m + 70.0 m = 77.35 m above the ground. Therefore, the total distance that the stone travels is 7.35 m (upward) + 77.35 m (downward) = 84.7 m (total).

CHAPTER 3

KINEMATICS IN TWO DIMENSIONS; VECTORS

OBJECTIVES

After studying the material of this chapter, the student should be able to:

- represent the magnitude and direction of a vector using a protractor and ruler.
- multiply or divide a vector quantity by a scalar quantity.
- use the methods of graphical analysis to determine the magnitude and direction of the vector resultant in problems involving vector addition or subtraction of two or more vector quantities. The graphical methods to be used are the parallelogram method and the tip to tail method.
- use the trigonometric component method to resolve a vector components in the x and y directions.
- use the trigonometric component method to determine the vector resultant in problems involving vector addition or subtraction of two or more vector quantities.
- use the kinematics equations of chapter two along with the vector component method of chapter three to solve problems involving two dimensional motion of projectiles.

KEY TERMS AND PHRASES

resultant vector is the arithmetic sum (or difference) of the magnitudes and the directions of two or more vectors.

tip to tail method is a graphical method used to determine the vector sum of two or more vectors. The two vectors are drawn to scale and then moved parallel to their original direction until the tail of one vector is at the tip of the next vector. Once all of the vectors are joined in this manner the resultant vector can be determined. The resultant is drawn from the tail of the first vector to the tip of the last vector. The angle of the resultant above (or below) the x axis is determined by using a protractor.

parallelogram method is a graphical method useful if two vectors are to be added. The two vectors are drawn to scale and joined at the tails. Dotted lines are then drawn from the tip of each vector parallel to the other vector. The finished diagram is a parallelogram. The resultant is along the diagonal of the parallelogram and extends from the point where the tails of the original vectors touch to the point where the dotted lines cross. The angle of the resultant above (or below) the x axis is determined by using a protractor.

vector component method is used to replace each vector with components in the x and y directions. The arithmetic sum of the x components (ΣX) and y components (ΣY) are then determined. Since ΣX and ΣY are at right angles, the Pythagorean theorem can be used to determine the magnitude of the resultant. The definition of the tangent of an angle can be used to determine the angle of the resultant above (or below) the x axis.

relative velocity refers to the velocity of an object with respect to a particular frame of reference.

projectile motion is the motion of an object fired (or thrown) at an angle θ with the horizontal. The only force acting on the object during its motion is gravity.

SUMMARY OF MATHEMATICAL FORMULAS

projectile motion equations	$v_{yo} = v_o \sin\theta$	Initial vertical component of a projectile's velocity where θ is measured from the horizontal direction.
	$v_{xo} = v_o \cos\theta$	Initial horizontal component of a projectile's velocity. Again, θ is measured from the horizontal direction.
	$y = v_{yo}\, t - ½\, g\, t^2$	Vertical component of an object's position (y) as related to its initial speed (v_{yo}), gravitational acceleration (g), and time of motion (t).
	$v_y^2 = v_{yo}^2 - 2\, g\, y$	Vertical component of an object's velocity (v_y) as related to its initial speed (v_{yo}), gravitational acceleration (g), and position (y).
	$v_y = v_{yo} - g\, t$	Vertical component of an object's velocity (v_y) as related to its initial speed (v_{yo}), gravitational acceleration (g), and time (t).
	$v_x = v_{xo}$	Horizontal component of velocity (v_x) remains constant during the projectile's flight.
	$x = x_o + v_{xo}\, t$	horizontal component of an object's position (x) as related to the initial horizontal speed (v_{xo}), and time (t)

CONCEPT SUMMARY

Addition or Subtraction of Vectors - Graphical Methods

If two vectors are in the same direction and along the same line, the **resultant vector** will be the arithmetic sum of the magnitudes and the direction will be in the direction of the original vectors. If the vectors are in opposite directions and along the same line, the resultant vector will have a magnitude equal to the difference in the magnitudes of the original vectors and the direction of the resultant will be in the direction of the vector which has the greater magnitude.

If the vectors are at some angle to each other than 0° or 180°, then special methods must be used to determine the magnitude and direction of the resultant. Two graphical methods used are the 1) tip to tail method and 2) parallelogram method.

The graphical methods involve the use of a ruler and a protractor. An appropriate scale factor must be used to represent the vectors. For example, a velocity vector which has a magnitude of 3.0 m/s and directed due east can be represented by a line 3.0 cm long directed along the +x axis. 3.0 cm +x

Tip to Tail Method

The tip to tail method can be conveniently used for the addition or subtraction of two or more vectors. The method consists of moving the vectors parallel to their original direction until the tail of one vector is at the tip of the next vector. Once all of the vectors are joined in this manner the resultant vector can be determined. The resultant is drawn from the tail of the first vector to the tip of the last vector. For example, the resultant velocity relative to the bank of a river for a person swimming at 3.0 m/s downstream in a river with a current of 1.0 m/s can be determined as shown in the figure. The magnitude of the resultant can be determined by measuring the length of the vector and multiplying by the scale factor. The scale used in the diagram is 1.0 cm = 1.0 m/s.

3.0 m/s

1.0 m/s

4.0 m/s

Using a ruler it can be determined that the resultant has a length of 4.0 cm. Therefore, the magnitude is (4.0 cm) x (1.0 m/s)/(1.0 cm) = 4.0 m/s. The direction of the resultant vector is downstream.

Parallelogram Method

The parallelogram method is useful if two vectors are to be added. If more than two vectors are involved, the method becomes cumbersome and an alternate method should be used.

The two vectors are drawn to scale and joined at the tails. Dotted lines are then drawn from the tip of each vector parallel to the other vector. A protractor can be used to ensure that the lines are drawn parallel. The finished diagram is a parallelogram. The resultant is along the diagonal of the parallelogram and extends from the point where the tails of the original vectors touch to the point where the dotted lines cross. As in the tip to tail method, the resultant is determined by measuring the length of the resultant and multiplying by the scale factor used to represent the vectors. The angle of the resultant above (or below) the x axis is determined by using a protractor.

For example, determine the sum of the following vectors: $\vec{v}_1$ = 3.0 m/s due east, $\vec{v}_2$ = 4.0 m/s due north.

step 1. Draw a diagram representing each with the tail of each joined at a point. Scale: let 1.0 cm = 1.0 m/s.

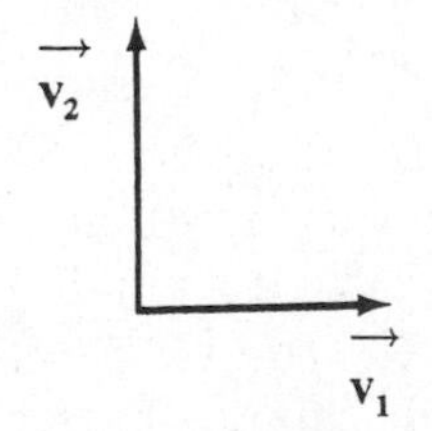

step 2. Complete the parallelogram. Draw the resultant vector across the diagonal. Determine the magnitude and direction of the resultant vector.

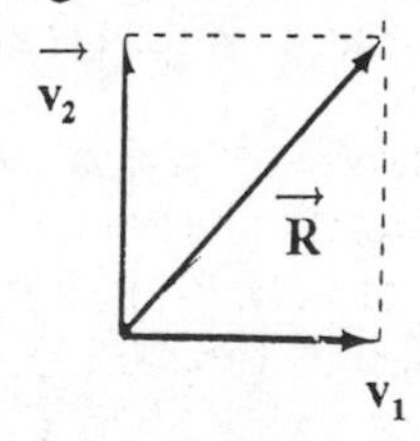

Using a ruler, it can be determined that the resultant has a length of 5.0 cm. Therefore, the magnitude of the resultant is (5.0 cm) x (1.0 m/s)/(1.0 cm) = 5.0 m/s. Using a protractor, the angle θ can be determined to be 53° north of east. Therefore, $\vec{v}_R$ = 5.0 m/s ∠53° N of E.

Subtraction of Vectors

The negative of a vector is a vector of the same magnitude but in the opposite direction. Thus, if vector $\vec{v}$ = 5 m/s due east, then $-\vec{v}$ = 5 m/s due west.

In order to subtract one vector from another, rewrite the problem so that the rules of vector addition can be applied. For example, $\vec{A} - \vec{B}$ can be rewritten as $\vec{A} + (-\vec{B})$. Determine the magnitude and direction of $-\vec{B}$ and apply the rules of vector addition to solve for the resultant vector.

Multiplication of a Vector by a Scalar

The product of a vector times a scalar has the same direction as the vector and a magnitude equal to the product of the magnitude of the scalar times the magnitude of the vector. For example, if c is a scalar while $\vec{V}$ is a vector, then the product has a magnitude cV and the same direction as $\vec{V}$. If c is a negative scalar, the magnitude of the resultant is still $c\vec{V}$ but the direction of the resultant is directly opposite that of $\vec{V}$.

Analytic Method for Adding Vectors

In the trigonometric component method, each of the original vectors is expressed as the vector sum of two other vectors. The two vectors are chosen to be in directions which are perpendicular to another one. At first, this method may seem long and tedious. However, with some practice this method is by far the most useful for this course.

For problems involving the addition of two or more vectors lying in the x-y plane, the method reduces to the following steps:

Step 1. Resolve each vector into x and y components.

If the angle is measured from the x axis to the vector, then the x component is equal to the product of the magnitude of the vector and the cosine of the angle. The y component is equal to the product of the magnitude of the vector and the sine of the angle. For example, for

$\vec{v}$ = 5.0 m/s ∠30° N of E

$v_x = v \cos \theta$ = (5.0 m/s)(cos 30°) = 4.3 m/s

$v_y = v \sin \theta$ = (5.0 m/s)(sin 30°) = 2.5 m/s

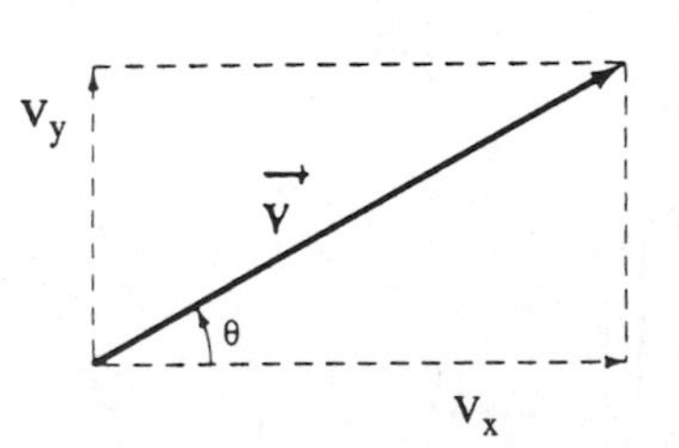

Step 2. Sign Convention

Assign a positive value to the magnitude if the component is in the +x or +y direction and a negative value if the component is in the -x or -y direction. Thus, for the vector used as an example in Step 1, v_x = + 4.3 m/s and v_y = + 2.5 m/s

Step 3. Reduce the problem to the sum of two vectors.

Determine the sum of the x components (ΣX) and the sum of the sum of the y components (ΣY), where Σ is the upper case Greek letter sigma which is designated to mean "the sum of." Since the x components are along the same line, their magnitudes are added arithmetically. This is also true for the y components.

Step 4. Determine the magnitude and direction of the resultant.

Since ΣX and ΣY are at right angles, the Pythagorean theorem can be used to determine the magnitude of the resultant. The definition of the tangent of an angle can be used to determine the direction of the resultant.

EXAMPLE PROBLEM 1. A man walks 100 meters due east, then 100 meters∠60.0° north of east. Determine the magnitude and direction of the resultant displacement using the a) tip to tail method, b) parallelogram method, and c) vector component method.

Part a. Step 1.
Displacement is a vector quantity. Write each vector in vector notation.

Solution: (Sections 3-2 and 3-4)

$\vec{D}_1$ = 100 m ∠due east

$\vec{D}_2$ = 100 m ∠60.0° north of east

Part a. Step 2.

Draw an accurate diagram locating each vector. Let the scale factor used in the drawing be 1.0 cm ≈ 50 m. A protractor must be used to measure the angles.

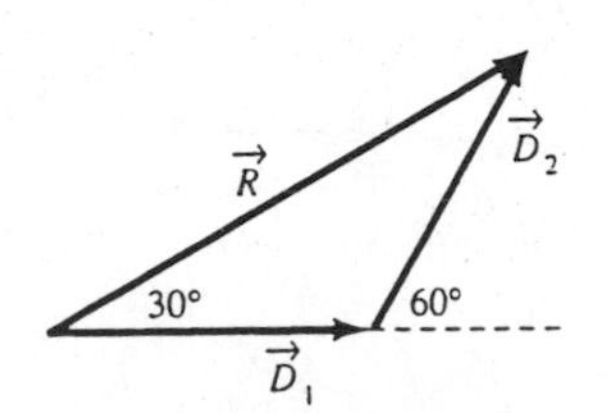

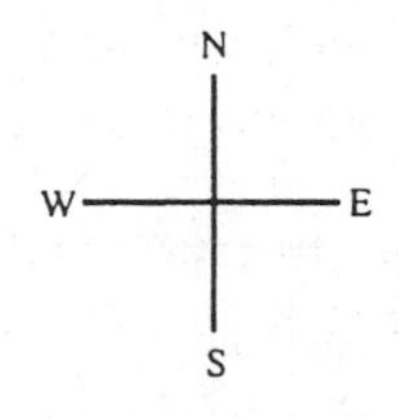

Part a. Step 3.

Determine the magnitude and direction of the resultant vector using the tip to tail method.

This method consists in moving one of the vectors parallel to its original direction until its tail is at the tip of the other vector. The resultant is drawn from the unattached tail of the second vector to the unattached head of the first.

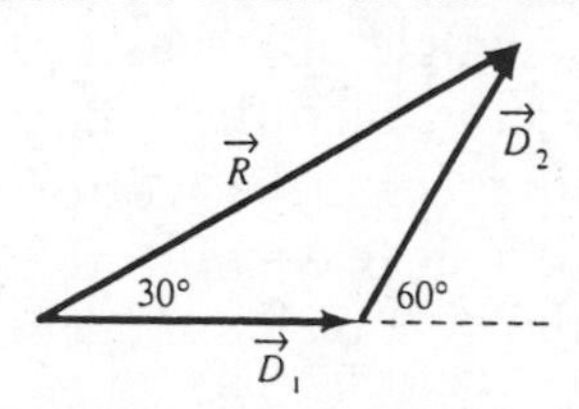

Based on careful measurement with a ruler, the magnitude of the resultant is approximately 3.4 cm. Using the scale factor, (3.4 cm)(50 m/1 cm) ≈ 170 m, the magnitude in meters can be determined. The direction of the resultant is determined by use of a protractor and is 30° N of E. Thus $\vec{\mathbf{R}}$ ≈ 170 m ∠30° N of E.

Part b. Step 1.

Draw the vectors and complete the parallelogram. Note: in the diagram shown at the right the scale factor is 1.0 cm ≈ 43 m.

The two vectors are drawn tail to tail. In order to complete the parallelogram, the dotted lines must be drawn parallel to the opposite sides. A protractor should be used to ensure that this is done properly.

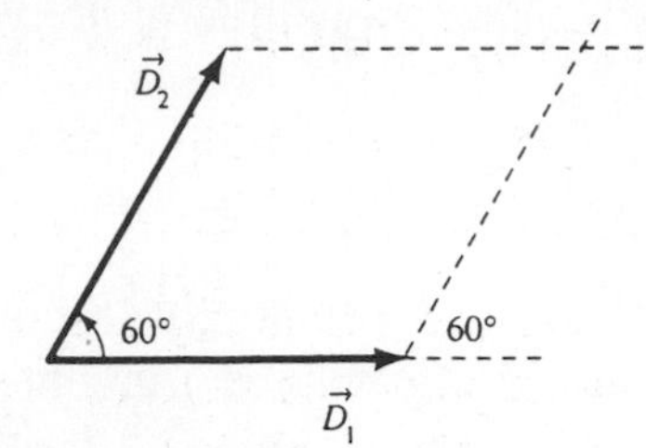

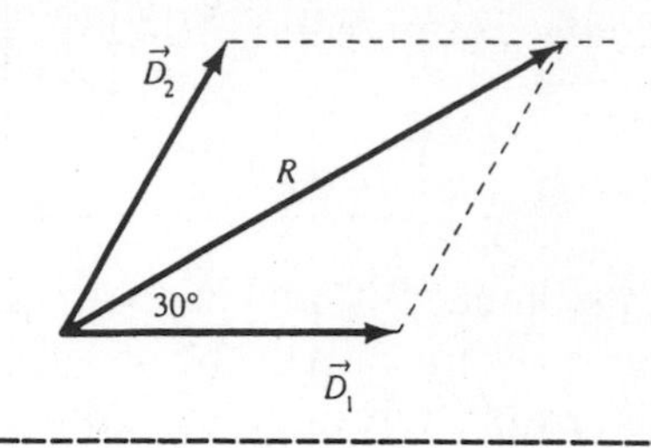

Part b. Step 2.

Use a protractor and a ruler to determine the magnitude and direction of the resultant. Note: remember to multiply by the scale factor.

The resultant is along the diagonal of the parallelogram. The direction is from the point where the tails are joined to the point where the dotted lines cross. As in the tip to tail method, the resultant is determined by measuring the length and multiplying by the scale factor. The direction is determined by using a protractor.

The length of R is measured to be approximately 4.0 cm; therefore, (4.0 cm)(43 m/1 cm) ≈ 170 m. Using a protractor, the angle is measured to be 30°. Thus, $\vec{\mathbf{R}} \approx 170\text{ m } \angle 30°$ N of E.

Vector Component Method: This method does not require an accurate diagram as part of the solution. However, a diagram reflecting the relative magnitudes and directions of the vectors is helpful.

Part c. Step 1.

Make a drawing and resolve each vector into x and y components.

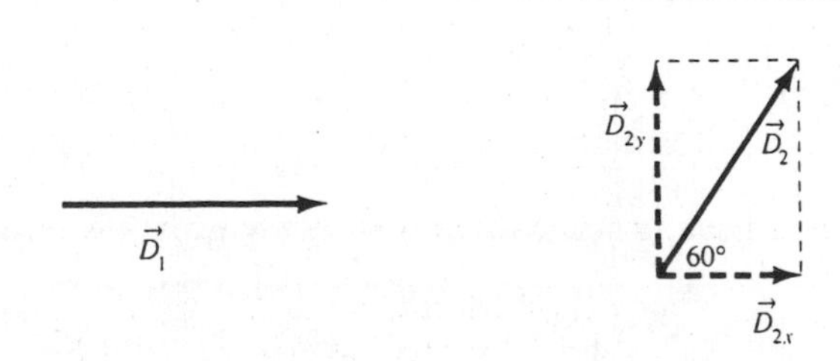

Part c. Step 2.

Use trigonometry to determine the magnitude of each component.

If the angle is measured from the east-west axis (x axis) to the vector, then the x component is equal to the product of the magnitude of the vector and the cosine of the angle. The y component is equal to the product of the magnitude of the vector and the sine of the angle.

x components

$\vec{\mathbf{D}}_{1x} = (100\text{ m})(\cos 0°) = +100\text{ m}$ $\vec{\mathbf{D}}_{2x} = (100\text{ m})(\cos 60.0°) = +50.0\text{ m}$

y components

$\vec{\mathbf{D}}_{1y} = (100\text{ m})(\sin 0°) = 0$ $\vec{\mathbf{D}}_{2y} = (100\text{ m})(\sin 60.0°) = +86.6\text{ m}$

Part c. Step 3. Determine the vector sum of the x components ($\Sigma\vec{X}$) and y components ($\Sigma\vec{Y}$).	The x components are in the +x direction; therefore, both are assigned positive values. The y-component of D_2 is in the +y direction; therefore, it has been assigned a positive value. $\Sigma\vec{X} = \vec{D}_{1x} + \vec{D}_{2x} = 100\ m + 50.0\ m = 150\ m$ $\Sigma\vec{Y} = \vec{D}_{1y} + \vec{D}_{2y} = 0\ m + 86.6\ m = 86.6\ m$
Part c. Step 4. Complete the parallelogram using ΣX and ΣY as the vectors to be added. Solve for the magnitude and direction of the resultant vector.	Since ΣX and ΣY are at right angles, the Pythagorean theorem and trigonometry can be used to determine the magnitude and direction of the resultant. $R = [(\Sigma X)^2 + (\Sigma Y)^2]^{1/2} = [(150\ m)^2 + (86.6\ m)^2]^{1/2} = 173\ m$ $\tan\theta = (86.6\ m)/(150\ m) = 0.577$; $\theta = 30.0^\circ$ N of E Thus, $\vec{R} = 173\ m\ \angle 30.0^\circ$ N of E.

TEXTBOOK QUESTION 1. One car travels due east at 40 km/h, and a second car travels due north at 40 km/h. Are their velocities equal? Explain.

ANSWER: Velocity is a vector quantity, i.e., it has both magnitude and direction. The two cars have the equal speed but they are traveling in different directions. Therefore, their velocities are not equal.

TEXTBOOK QUESTION 4. During baseball practice, a batter hits a very high fly ball, and then runs in a straight line and catches it. Which had the greater displacement, the player or the ball?

ANSWER: The distance traveled by the ball is greater than the distance traveled by the player. But distance is a scalar quantity not a vector quantity. Displacement is a vector quantity and is the same for each. The displacement vector is the distance and straight line direction between the starting point and the ending point.

TEXTBOOK QUESTION 6. Two vectors have length V_1 = 3.5 km and V_2 = 4.0 km. What are the maximum and minimum magnitudes of their vector sum?

ANSWER: The maximum magnitude occurs if the two vectors are in the same direction. The resultant velocity is the arithmetic sum of their magnitudes, i.e., V_R = 3.5 km + 4.0 km = 7.5 km.

The minimum magnitude occurs if the two vectors are in opposite directions. The resultant velocity is the difference of their magnitudes, i.e., V_R = 4.0 km - 3.5 km = 0.5 km in the direction of the vector which has the greater magnitude.

Relative Velocity

Relative velocity refers to the velocity of an object with respect to a particular frame of reference. As in chapter 2, the reference frame is usually specified by using Cartesian coordinates, i.e., x, y, and z axes, relative to which the position and/or motion of an object can be determined. As stated in the textbook, the velocity of an object relative to one frame of reference can be found by vector addition if its velocity relative to a second frame of reference and the relative velocity of the two reference frames are known.

EXAMPLE PROBLEM 2. The current in a river is 1.00 m/s. A woman swims 600 m downstream and then back to her starting point without stopping. If she can swim 2.00 m/s in still water, determine the time required for the round trip.

Part a. Step 1.

Determine her velocity relative to the river bank as she swims downstream.

Solution: (Section 3-8)

Let $\vec{\mathbf{v}}_{pw}$ = velocity of the person relative to the river water

$\vec{\mathbf{v}}_{ws}$ = velocity of the water relative to the shore of the river, i.e., the river current

$\vec{\mathbf{v}}_{ps}$ = velocity of the person relative to the shore of the river

The tip to tail method can be used to solve for $\mathbf{v}_{ps}$. The vectors are along the same line; therefore, the resultant vector is the arithmetic sum of the magnitudes.

$$\vec{\mathbf{v}}_{ps} = \vec{\mathbf{v}}_{pw} + \vec{\mathbf{v}}_{ws}$$

$$\vec{\mathbf{v}}_{ps} = 2.00 \text{ m/s} + 1.00 \text{ m/s}$$

$$\vec{\mathbf{v}}_{ps} = 3.00 \text{ m/s (downstream)}$$

2.00 m/s
1.00 m/s
3.00 m/s

Part a. Step 2.

Determine the time to swim downstream.

time = displacement/resultant velocity

$$t = \vec{\mathbf{D}}/\vec{\mathbf{v}}_{ps} = (600 \text{ m})/(3.00 \text{ m/s}) = 200 \text{ s}$$

Part a. Step 3.

Determine the time required to swim back to the starting point.

The tip to tail method can again be used to solve for $\vec{\mathbf{v}}_{ps}$. The woman is now swimming against the current. Thus

$$\vec{\mathbf{v}}_{ps} = \vec{\mathbf{v}}_{pw} - \vec{\mathbf{v}}_{ws}$$

$$= \vec{\mathbf{v}}_{pw} + (-\vec{\mathbf{v}}_{ws})$$

$$\vec{\mathbf{v}}_{ps} = 2.00 \text{ m/s} + (-1.00 \text{ m/s}) = 1.00 \text{ m/s (upstream)}$$

$\vec{\mathbf{v}}_{ps}$ $\vec{\mathbf{v}}_{ps}$ $\vec{\mathbf{v}}_{ws}$

time = displacement/resultant velocity

$$t = \vec{\mathbf{D}}/\vec{\mathbf{v}}_{ps} = (600 \text{ m})/(1.00 \text{ m/s}) = 600 \text{ s}$$

Part a. Step 4.	$t_{total} = t_{downstream} + t_{upstream}$
Determine the total time for the round trip.	t_{total} = 200 s + 600 s = 800 s

Projectile Motion

Projectile motion is the motion of an object fired at an angle θ with the horizontal. This motion can be discussed by analyzing the horizontal component of the object's motion independent of the vertical component of motion. If air resistance is negligible, then the horizontal component of motion does not change; thus $a_x = 0$ and $v_x = v_{xo}$ = constant.

The vertical component of motion is affected by gravity and is described by the equations for an object in free fall discussed in the textbook in chapter 2. The following equations are used to describe the motion of a projectile.

vertical component of motion | horizontal component of motion

$v_y = v_{yo} - gt$ | $x = x_o + v_{xo} t$

$y = y_o + v_{yo} t - \frac{1}{2} g t^2$ | where $v_x = v_{xo}$

$v_y^2 = v_{yo}^2 - 2 g y$

v_y is the vertical component of velocity at time t.

v_{yo} is the initial vertical component of velocity, $v_{yo} = v_o \sin \theta$.

g is the acceleration due to gravity, g = 9.8 m/s².

t is the time interval of the motion, t_o = 0 s.

y is the vertical displacement from the release point. The origin of the coordinate system is the release point (y_o) with the upward direction taken as positive.

x is the horizontal displacement from the release point. The release point (x_o) is usually taken to be the zero point of motion in the horizontal direction, i.e., $x_o = 0$.

v_{xo} is the initial horizontal component of velocity. Assuming air resistance to be negligible, the horizontal component of velocity does not change during the motion. Therefore, $v_x = v_{xo} = v_o \cos \theta$.

TEXTBOOK QUESTION 19. A projectile is launched at an angle of 30° to the horizontal with a speed of 30 m/s. How does the horizontal component of its velocity 1.0 s after launch compare with its horizontal component of velocity 2.0 s after launch?

ANSWER: The motion of a projectile at any point in its motion is the vector sum of the horizontal component and the vertical component of its velocity. As a projectile travels upward the magnitude of the vertical component decreases, reaching zero at maximum height. On the way down, the vertical component of velocity again increases. However, in the absence of air resistance, the magnitude of horizontal component does **not** change. The magnitude of horizontal component $v_x = v_{xo}$ = (30 m/s)(cos 30°) = 26 m/s at each point in the motion.

EXAMPLE PROBLEM 3. A stone is thrown horizontally outward from the top of a bridge. The stone is released 19.6 meters above the street below. The initial velocity of the stone is 10.0 m/s. Determine the a) total time that the stone is in the air and b) magnitude and direction of the velocity of the projectile "just" before it strikes the street.

	Solution: (Sections 3-5 and 3-6)
Part a. Step 1. Draw an accurate diagram showing the trajectory of the projectile.	19.6 m
Part a. Step 2. Determine the initial horizontal, and vertical components of velocity.	Since the object is thrown horizontally, the initial vertical component (v_{yo}) is zero. The initial horizontal component (v_{xo}) is 10.0 m/s. Note: the vertical component of motion is independent of the horizontal motion.
Part a. Step 3. Complete a data table using information both given and implied in the problem.	Using the technique discussed in the textbook, the upward direction will be designated as the positive. The top of the cliff is the zero point of position, i.e. $y_o = 0$. Therefore, the bottom, of the cliff is at y = -19.6 m. $g = 9.8\ m/s^2$ $v_{yo} = 0$ $t = ?$ $y = 19.6$ m $v_y = ?$ $x = ?$ $v_{xo} = 10.0$ m/s
Part a. Step 4. Determine the total time that the stone is in the air.	$y = v_{yo}\,t - \frac{1}{2}\,g\,t^2$ $-19.6\ m = (0\ m/s)\ t - \frac{1}{2}\ (9.8\ m/s^2)\ t^2$ $t^2 = 2(-19.6\ m)/(-9.8\ m/s^2) = 4.0\ s^2$ Either t = 2.0 s or - 2.0 s. Since the time of flight cannot be negative, the answer is t = 2.0 s.
Part b. Step 1. Determine the vertical velocity just before it hits the street.	$v_y = v_{yo} - g\,t$ $= 0\ m/s - (9.8\ m/s^2)(2.0\ s)$ $v_y = -\ 19.6\ m/s$

Part b. Step 2.

Determine the magnitude and direction of the velocity of the projectile just before it strikes the street.

The horizontal velocity remains constant; therefore, both components of motion are now known. Use the Pythagorean theorem and trigonometry to determine the magnitude and direction of the resultant velocity (v_R).

$$v_R = [(v_y) + (v_{xo})]^{1/2}$$

$$= [(-19.6 \text{ m/s})^2 + (10.0 \text{ m/s})^2]^{1/2}$$

$$v_R = 22.0 \text{ m/s}$$

$$\tan \theta = v_y/v_{xo} = (-19.6 \text{ m/s})/(10.0 \text{ m/s}) = 1.96; \quad \theta = -63.0^\circ$$

$$\vec{v}_R = 22.0 \text{ m/s } \angle 63.0^\circ \text{ below the horizontal}$$

EXAMPLE PROBLEM 4. A projectile is fired with an initial speed of 196 m/s at an angle of 30.0° above the horizontal from the top of a cliff 98.0 m high. Determine the a) time to reach maximum height, b) maximum height above the base of the cliff reached by the projectile, c) total time it is in the air, and d) horizontal range of the projectile.

Part a. Step 1.

Draw an accurate diagram showing the trajectory of the projectile.

Solution: (Sections 3-5 and 3-6)

Part a. Step 2.

Determine v_{yo} and v_{xo}.

$$v_{yo} = v_o \sin 30.0^\circ = (196 \text{ m/s})(0.500) = 98.0 \text{ m/s}$$

$$v_{xo} = v_o \cos 30^\circ = (196 \text{ m/s})(0.866) = 170 \text{ m/s}$$

Part a. Step 3.

Complete a data table using information both given and implied.

$v_{yo} = 98.0$ m/s

$t = ?$ (to reach max height)

$y = ?$ (at max height)

$v_y = 0$ (at maximum height)

$g = 9.80 \text{ m/s}^2$

$v_{xo} = 170$ m/s

$x = ?$

Part a. Step 4.

Determine the time to reach maximum height.

$$v_y = v_{yo} - gt$$

$$0 \text{ m/s} = 98.0 \text{ m/s} - (9.80 \text{ m/s}^2)\, t$$

$$t = 10.0 \text{ s}$$

Part b. Step 1. Determine the maximum height above the cliff reached by the projectile Now determine the maximum height above the ground reached by the projectile.	$y = -\frac{1}{2} g t^2 + v_{yo} t$ $= -\frac{1}{2} (9.80 \text{ m/s}^2)(10.0 \text{ s})^2 + (98.0 \text{ m/s})(10.0 \text{ s})$ $= -490 \text{ m} + 980 \text{ m}$ $y = 490 \text{ m}$ The total height above the base of the cliff can now be determined: $y = 490 \text{ m} + 98.0 \text{ m} = 588 \text{ m}$
Part c. Step 1. Complete a data table and determine the time required for the projectile to fall from maximum height to the bottom of the cliff.	Using the technique discussed in the textbook, the upward direction is designated as positive. The top of the trajectory is the zero point of position, i.e., $y_o = 0$. Therefore, the bottom, of the cliff is at $y = -588$ m. $g = 9.8 \text{ m/s}^2$ $v_{yo} = 0$ (at maximum height) $t = ?$ $y = -588 \text{ m}$ $y = v_{yo} t - \frac{1}{2} g t^2$ $-588 \text{ m} = (0 \text{ m/s}) t - \frac{1}{2} (9.80 \text{ m/s}^2) t^2$ $t^2 = 2(-588 \text{ m})/(-9.80 \text{ m/s}^2) = 120 \text{ s}^2$ $t = 11.0 \text{ s}$
Part c. Step 2. Determine the total time the projectile is in the air.	total time = 10.0 s + 11.0 s = 21.0 s
Part d. Step 1. Use the equation for horizontal motion to determine the horizontal range.	$x = v_{xo} t = (170 \text{ m/s})(21.0 \text{ s})$ $x = 3570 \text{ m}$

PROBLEM SOLVING SKILLS

For problems involving vector addition or subtraction:

1. Use a protractor and ruler to accurately represent each vector involved in the problem. Make sure to use an appropriate scale factor in representing the vector.
2. Choose an appropriate method to solve the problem. If a graphical method is used, be sure to measure both the magnitude and direction of the resultant.

3. If the trigonometric component method is used, you must
   a) break each vector into x and y components.
   b) use the sign convention and assign a positive sign or a negative sign to the magnitude.
   c) determine the sum of the x components and repeat for the sum of the y components.
4) Use the Pythagorean theorem and simple trigonometry to solve for the magnitude direction of the resultant.

For problems involving projectile motion:

1. Draw an accurate diagram showing the trajectory of the projectile.
2. Use the trigonometric component method to determine v_{xo} and v_{yo}.
3. Complete a data table using information both given and implied in the wording of the problem.
4. As in the free fall problems of chapter 2, use the appropriate sign convention depending on whether the object was initially moving upward or downward.
5. Memorize the formulas for projectile motion. It is also necessary to memorize the meaning of each symbol in each formula. Using the data from the completed data table, determine which formula or combination of formulas must be used to solve the problem.

SOLUTIONS TO SELECTED TEXTBOOK PROBLEMS

TEXTBOOK PROBLEM 9. An airplane is traveling 735 km/h in a direction 41.5° west of north (Fig. 3-31). (a) Find the components of the velocity vector in the northerly and westerly directions. (b) How far north and how far west has the plane traveled after 3.00 h?

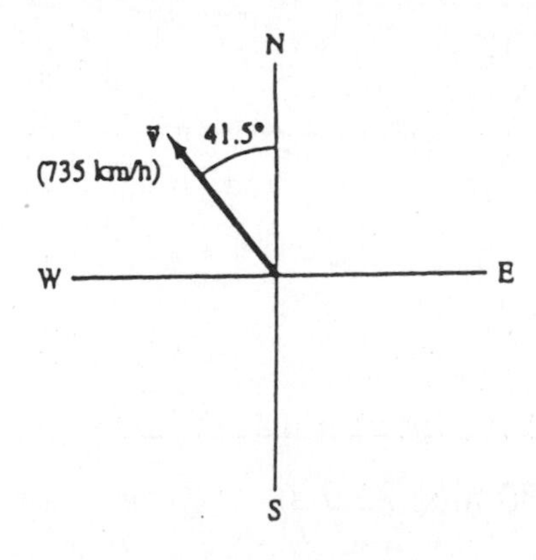

Part a. Step 1. Determine the angle θ as measured north of west.	Solution: (Section 3-4) θ = 90.0° - 41.5° W of N = 48.5° N of W.
Part a. Step 2. Use trigonometry to determine the north component and the west component	v_{north} = (735 km/h) sin 48.5° = 550 km/h v_{west} = (735 km/h) cos 48.5° = 487 km/h

Part b. Step 1. Determine the distance traveled in 3.00 h.	d_{north} = (550 km/h)(3.00 h) = 1650 km d_{west} = (487 km/h)(3.00 h) = 1460 km

TEXTBOOK PROBLEM 10. Three vectors are shown in Fig. 3-32. Their magnitudes are given in arbitrary units. Determine the sum of the three vectors. Give the resultant in terms of (a) components, and (b) magnitude and angle with the x axis.

Part a. Step 1.

Determine the x and y component of each vector and then determine ΣX and ΣY. Remember to apply the sign convention.

Solution: (Section 3-4)

A_x = 44.0 cos 28.0° = + 38.8	A_y = 44.0 sin 28.0° = + 20.6
B_x = 26.5 cos 56.0° = - 14.8	B_y = 26.5 sin 56.0° = + 22.0
C_x = 31.0 cos 90.0° = 0.00	C_y = 31.0 sin 90.0° = - 31.0
ΣX = + 24.0	ΣY = + 11.6

Part a. Step 2.

Draw a vector diagram showing ΣX and ΣY. Complete the parallelogram and draw in the resultant.

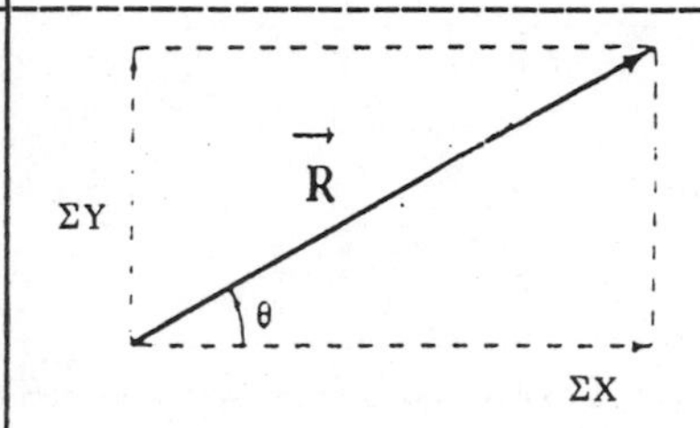

Part a. Step 3.

Use the Pythagorean theorem and trigonometry to determine the magnitude and direction of the resultant.

$R^2 = (\Sigma X)^2 + (\Sigma y)^2 = (+ 24.0)^2 + (+ 11.6)^2 = 711$

R = 26.7

tan θ = (+ 11.6)/(+ 24.0) = 0.488; therefore, θ = 25.8° N of E

$\vec{\mathbf{R}}$ = 26.7 ∠25.8° N of E

TEXTBOOK PROBLEM 21. A ball is thrown horizontally from the roof of a building 45.0 m tall and lands 24.0 m from the base. What was the ball's initial speed.

Part a. Step 1.

Draw an accurate diagram showing the trajectory of the projectile.

Solution: (Sections 3-5 and 3-6)

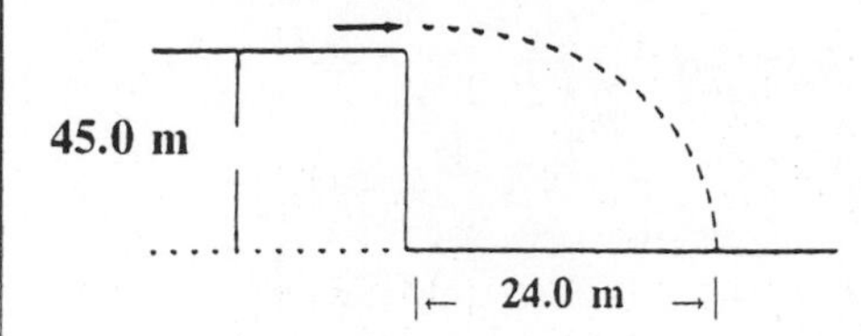

Part a. Step 2.. Complete a data table using information both given and implied in the problem.	Using the technique discussed in the textbook, the upward direction will be designated as positive. The top of the building is the zero point of position, i.e. $y_o = 0$. Therefore, the street below is at $y = -45.0$ m. $g = 9.8\ m/s^2$ $v_{yo} = 0$ $t = ?$ $y = -45.0$ m $v_y = ?$ $x = 24.0$ m $x_o = 0$ $v_{xo} = ?$
Part a. Step 3. Determine the total time that the stone is in the air.	$y = v_{yo}\, t - \frac{1}{2}\, g\, t^2$ $-45.0\ m = (0\ m/s)\ t - \frac{1}{2}\ (9.8\ m/s^2)\ t^2$ $t^2 = 2(-45.0\ m)/(-9.8\ m/s^2) = 9.18\ s^2$ Either t = 3.03 s or - 3.03 s. Since the time of flight cannot be negative, the answer is t = 3.03 s.
Part a. Step 4. Determine the ball's initial speed.	The horizontal velocity remains constant, i.e., $v_{xo} = v_x$. $x = v_{xo}\, t$ $24.0\ m = v_{xo}\ (3.03\ s)$ $v_{xo} = 7.92$ m/s

TEXTBOOK PROBLEM 24. An athlete executing a long jump leaves the ground at a 28.0° angle and travels 7.80 m. (a) What was the takeoff speed? (b) If this speed was increased by just 5.0 percent, how much longer would the jump be?

Part a. Step 1. Draw an accurate diagram showing the trajectory of the athlete.	Solution: (Sections 3-5 and 3-6) 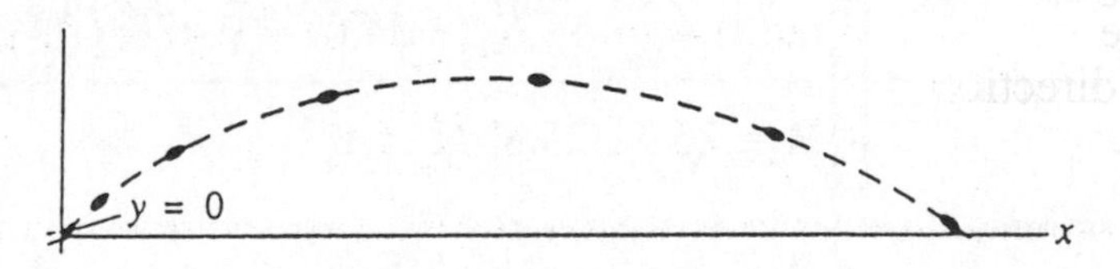
Part a. Step 2. Complete a data table using information both given and implied.	Since the initial motion is upward, the upward direction will be designated as the positive direction. Also, the athlete starts at ground level and returns to ground level; therefore, $y = y_o = 0$ m. $g = 9.80\ m/s^2$ $t = ?$ $y = 0$ m $y_o = 0$ m $x = 7.80$ m $x_o = 0$ $v_y = ?$ $v_{yo} = v_o \sin 28.0° = (0.469)\ v_o$ $v_{xo} = v_o \cos 28.0° = (0.883)\ v_o$

Part a. Step 3. Write an equation for the horizontal distance in terms of the horizontal speed and time.	$x = v_{xo}\ t$ $7.80\ m = (0.883)\ v_o\ t$ $v_o\ t = 8.83\ m$
Part a. Step 4. Determine the time in air.	$y = v_{yo}\ t\ -\ ½\ g\ t^2$ $0 = (0.469)\ v_o\ t - (4.90\ m/s^2)\ t^2$ but $v_o\ t = 8.83\ m$ $0 = (0.469)(8.83\ m) - (4.90\ m/s^2)\ t^2$ $t^2 = 0.846\ s^2$ and $t = 0.920\ s$
Part a. Step 5. Combine the results of steps 2 and 3 to determine the initial speed.	$v_o\ t = 8.83\ m$ but $t = 0.920\ s$ $v_o\ (0.920\ s) = 8.83\ m$ $v_o = 9.60\ m/s$
Part b. Step 1. Determine the value of the velocity.	As a result of the 5 percent increase, the new velocity would be 1.05 greater than in part a. $v_o = (1.05)(9.60\ m/s) = 10.1\ m/s$
Part b. Step 2. Complete a data table using the new information.	$g = 9.80\ m/s^2$ $t = ?$ $y = 0\ m$ $y_o = 0\ m$ $x = ?\ m$ $x_o = 0$ $v_y = ?$ $v_{yo} = (10.1\ m/s)\ \sin 28.0° = 4.74\ m/s$ $v_{xo} = (10.1\ m/s)\ \cos 28.0° = 8.92\ m/s$
Part b. Step 3. Determine the time in air.	$y = v_{yo}\ t\ -\ ½\ g\ t^2$ $0 = (4.74\ m/s)\ t - (4.9\ m/s^2)\ t^2$ Using algebra gives either $t = 0.967\ s$ or $t = 0\ s$
Part b. Step 4. Determine the horizontal distance.	$x = v_{xo}\ t$ $x = (8.92\ m/s)(0.967\ s)$ $x = 8.60\ m$

Part b. Step 5.

Determine the increased distance.

8.60 m - 7.80 m = 0.80 m

Alternate Method. This problem is readily solved by using the range equation derived in the textbook. This special equation can only be used because the athlete returns to his original height above the ground.

Part a. Step 1.

Use the range equation to determine the take-off speed.

$R = v_o^2 \sin 2\theta/g$ and rearranging gives

$v_o^2 = R\, g/\sin 2\theta$ where $\sin 2\theta = \sin 2(28^\circ) = \sin 56^\circ = 0.829$

$v_o^2 = (7.80\text{ m})(9.8\text{ m/s}^2)/(0.829) = 92.2\text{ m}^2/\text{s}^2$

$v_o = 9.60\text{ m/s}$

Part b. Step 2.

Use the range equation to determine the increased distance.

$R = v_o^2 \sin 2\theta/g$

$R = [(1.05)(9.60\text{ m/s})]^2(0.829)/(9.8\text{ m/s}^2) = 8.60\text{ m}$

Increased distance: 8.60 m - 7.80 m = 0.80 m

TEXTBOOK PROBLEM 47. A swimmer is capable of swimming 0.45 m/s in still water. (a) If she aims her body directly across a 75 m wide river whose current is 0.40 m/s, how far downstream (from a point opposite her starting point) will she land? (b) How long will it take her to reach the other side?

Part a. Step 1.

Determine the woman's velocity relative to the riverbank ($\vec{v}_{ps}$).

Solution: (Section 3-8)

The woman's velocity relative to the water ($\vec{v}_{pw}$) is at right angles to the current ($\vec{v}_{ws}$). The Pythagorean theorem can be used to solve for the woman's velocity relative to the river bank $\vec{v}_{ps}$.

$v_{ps} = [(v_{pw})^2 + (v_{ws})^2]^{1/2}$

$\quad = [(0.45\text{ m/s})^2 + (0.40\text{ m/s})^2]^{1/2}$

$v_{ps} = 0.60\text{ m/s}$, $\tan\theta = (0.40\text{ m/s})/(0.45\text{ m/s}) = 0.89$, $\theta = 41.6^\circ$

$\vec{v}_{ps} = 0.60\text{ m/s}\ \angle 41.6^\circ$

Part a. Step 2.

Determine the distance downstream where she reaches the opposite bank of the river.

$\tan\theta = (D_{downstream})/(D_{across\ river})$

$\tan 41.6^\circ = (D_{downstream})/(75\text{ m})$

$D_{downstream} = (75\text{ m})(\tan 41.6^\circ) = (75\text{ m})(0.89) = 67\text{ m}$

Part b. Step 1.	
Determine the time required to cross the river.	time = displacement/resultant velocity $t = D/v_{ps}$ but as shown in the diagram, D is the actual distance traveled crossing the river. But $\cos 41.6° = (75\text{ m})/D$ $D = (75\text{ m})/(\cos 41.6°) = (75\text{ m})/(0.74.8) = 100\text{ m}$ $t = D/v_{ps} = (100\text{ m})/(0.60\text{ m/s}) = 170\text{ s}$

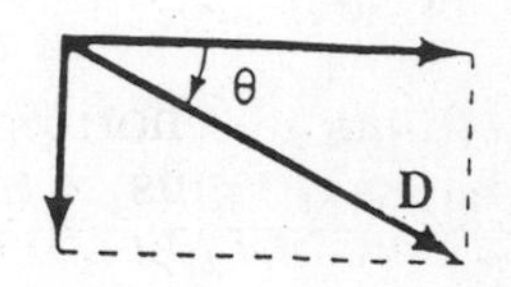

Alternate Method: Since the woman swims perpendicular to the current, the current does not affect the time for the trip across. An analogous situation is that of an airplane passenger walking from a window on the left side of the plane to a window on the right side of the plane. The time is the same if the plane is at rest or if it is moving at 450 miles per hour. Thus, time to cross river equals the distance across the river divided by the woman's velocity perpendicular to the river current.

$t = D/v_{pw} = (75\text{ m})/(0.45\text{ m/s}) = 170\text{ s}$

TEXTBOOK PROBLEM 62. When Babe Ruth hit a homer over a 7.5 m high right-field fence 95 m from home plate, roughly what was the minimum speed of the ball when it left the bat? Assume that the ball was hit 1.0 m above the ground and its path initially made a 38° angle with the ground.

Part a. Step 1.	Solution: (Sections 3-5 and 3-6)
Draw an accurate diagram showing the path of the ball.	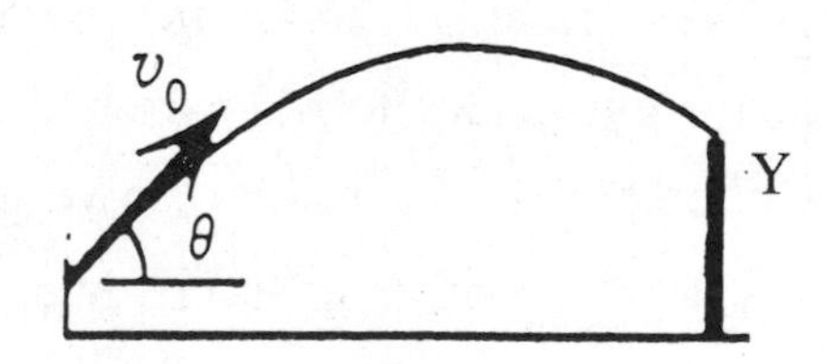

Part a. Step 2.				
Complete a data table using the information both given and implied.	$g = 9.8\text{ m/s}^2$	$t = ?$	$y = 7.5\text{ m}$	$y_0 = 1.0\text{ m}$
	$x = 95\text{ m}$	$x_o = 0$	$v_y = ?$	
	$v_{yo} = v_o \sin 38° = 0.62\, v_o$			
	$v_{xo} = v_o \cos 38° = 0.79\, v_o$			

Part a. Step 3. Determine the horizontal distance in terms of the initial velocity and time of flight.	$x = v_{xo}\, t$ $95\ m = 0 + 0.79\ v_o\, t$ $v_o\, t = 121\ m$
Part a. Step 4. Determine the time required for the ball to reach the right-field fence.	$y = y_o + v_{yo}\, t - \frac{1}{2} g\, t^2$ $7.5\ m = 1.0\ m + 0.62\ v_o\, t - (4.9\ m/s^2)\ t^2$ but $v_o\, t = 121\ m$ $6.5\ m = (0.62)(121\ m) - (4.9\ m/s^2)\ t^2$ $t^2 = (-68.5\ m)/(-4.9\ m/s^2) = 13.9\ s^2$ $t = 3.74\ s$
Part a. Step 5. Determine the ball's initial speed.	From Part a. Step 3. $v_o\, t = 121\ m$ but $t = 3.74\ s$ $v_o\ (3.74\ s) = 121\ m$ $v_o \approx 32\ m/s$

CHAPTER 4

DYNAMICS: NEWTON'S LAWS OF MOTION

OBJECTIVES

After studying the material of this chapter, the student should be able to:

- state Newton's three laws of motion and give examples that illustrate each law.
- explain what is meant by the term net force.
- use the methods of vector algebra to determine the net force acting on an object.
- define each of the following terms: mass, inertia, weight, and distinguish between mass and weight.
- identify the SI units for force, mass, and acceleration.
- draw an accurate free body diagram locating each of the forces acting on an object or a system of objects.
- use free body diagrams and Newton's laws of motion to solve word problems.

KEY TERMS AND PHRASES

dynamics is the study of the causes of motion.

net force refers to the vector sum of all of the forces acting on an object.

Newton's first law of motion is also known as Galileo's law of inertia, where **inertia** refers to the tendency of an object to resist any change in its state of motion. Inertia is measured by measuring an object's mass.

Newton's second law of motion refers to an object's motion when a net force does not equal zero. The net force (F) will cause an object to accelerate or decelerate. The rate of acceleration (a) is directly proportional to the magnitude of the net force and inversely proportional to the object's mass (m), i.e., $\vec{\mathbf{a}} = \Sigma\vec{\mathbf{F}}/m$ or $\Sigma\vec{\mathbf{F}} = m\,\vec{\mathbf{a}}$.

Newton's third law states that whenever one object exerts a force on a second object, the second object exerts an equal but opposite force on the first.

weight is a measure of the force of gravity on an object.

friction is a contact force that opposes the relative motion of two surfaces as they slide past each other. The frictional force depends on the coefficient of friction (μ) and the normal force (F_N).

coefficient of friction (μ) is a pure number without physical units and varies with the types of surfaces that are in contact. When the object is at rest, the **coefficient of static friction** (μ_s) is used to determine the magnitude of the frictional force just before the object starts to move. When the object is moving, the **coefficient of kinetic friction** (μ_k) is used.

normal force is a contact force that acts perpendicular to the common surface of contact.

SUMMARY OF MATHEMATICAL FORMULAS

Newton's first law	Net $\vec{\mathbf{F}} = 0$ or $\Sigma\vec{\mathbf{F}} = \mathbf{0}$	If the net force on an object equals zero, then the object's motion will not change, i.e., a = 0.
Newton's second law	$\vec{\mathbf{a}} = \Sigma\vec{\mathbf{F}}/m$ or $\Sigma\vec{\mathbf{F}} = m\,\vec{\mathbf{a}}$	An object's acceleration is related to the net force and the object's mass.
Newton's third law	$\vec{\mathbf{F}}_{12} = -\vec{\mathbf{F}}_{21}$	If object 1 exerts a force on object 2, then object 2 exerts an equal but opposite force on object 1.
weight	$\vec{\mathbf{w}} = m\,\vec{\mathbf{g}}$	An object's weight is directly proportional to its mass and gravitational acceleration.
force of friction	$F_{fr} = \mu\ F_N$	Frictional force depends on the coefficient of friction and the normal force.

CONCEPT SUMMARY

Newton's Laws of Motion

Dynamics is the study of the causes of motion. The basic causes of motion can be explained by using **Newton's three laws of motion.** The **first law** explains the motion of an object on which the **net force** equals zero. The net force refers to the vector sum of all of the forces acting on an object. The magnitude and direction of the net force is determined by using the methods of vector addition described in chapter 3 in the textbook. While the tip to tail and the parallelogram methods are useful, the method used most frequently in determining the resultant is the **vector component method.** If the net force equals zero, the object will tend to remain at rest, or if in motion, will remain in motion in a straight line at a constant speed, i.e., constant velocity.

Newton's first law of motion is also known as Galileo's law of inertia, where **inertia** refers to the tendency of an object to resist any change in its state of motion. Inertia is measured by measuring an object's mass.

Newton's second law of motion refers to an object's motion when a net force does not equal zero. The net force will cause an object to accelerate or decelerate. The rate of acceleration (a) is directly proportional to the magnitude of the net force and inversely proportional to the object's **mass** (m).

$\vec{\mathbf{a}} = \Sigma\vec{\mathbf{F}}/m$ or $\Sigma\vec{\mathbf{F}} = m\,\vec{\mathbf{a}}$

The SI unit of force is the **newton** (N) and for mass it is the kilogram (kg). A net force of one newton will cause a 1 kg object to accelerate at 1 m/s^2; thus 1 N = 1 kg m/s^2.

Newton's third law is the "action-reaction" law with which most students are familiar. However, it is necessary to be very careful in interpreting the meaning of this law. The action force and the reaction force are equal and opposite but do not act on the same object. A good way to remember the law is the statement given in the textbook: "Whenever one object exerts a force on a second object, the second object exerts an equal but opposite force on the first."

TEXTBOOK QUESTION 9. A stone hangs by a fine thread from the ceiling, and a section of the same thread dangles from the bottom of the stone (Fig. 4-36). If a person gives a sharp pull on the dangling thread, where is the thread likely to break: below the stone or above it? What if the person gives a slow and steady pull? Explain your answers.

ANSWER: The key to the answer lies in the concept of inertia. Inertia is the tendency of an object to resist any change in its state of motion. The stone has inertia and this inertia must be overcome in order to move the stone. If a person gives a SHARP pull to the dangling thread, the inertia of the stone resists any change in its motion. The result is that the stone does not move and the force exerted by the person is not transmitted to the section of thread above the stone. The tension in the section below the stone is much greater than above the stone and the lower section snaps.

If the person exerts a slow and steady pull, the inertia of the stone is overcome and the stone will begin to move slowly. The tension in the section of thread is due to the person. The tension in the section above the stone is due to the sum of the pull exerted by the person and the stone's weight. The tension in the section above the stone is greater than below the stone. As a result the section of thread above the stone breaks.

TEXTBOOK QUESTION 10. The force of gravity on a 2-kg rock is twice as great as that on a 1-kg rock. Why then doesn't the heavier rock fall faster?

ANSWER: According to Newton's second law of motion, an object's rate of acceleration is directly proportional to the magnitude of the object's weight but inversely proportional to the object's mass, i.e. $\vec{\mathbf{a}} = \Sigma\vec{\mathbf{F}}/m$. A 2 kg rock has twice as much weight as a 1 kg rock but it also has twice as much mass, i.e., inertia, and is twice as hard to accelerate. The ratio of the force to mass is the same for both objects and as a result the acceleration is the same for each.

TEXTBOOK QUESTION 17. When you stand still on the ground, how large a force does the ground exert on you? Why doesn't this force make you rise up into the air?

ANSWER: The downward force of your weight is balanced by an equal but opposite force exerted by the ground. The net force on your body is zero, your rate of acceleration is zero, and you remain motionless. As a result, you neither rise nor fall.

TEXTBOOK QUESTION 18. Whiplash sometimes results from an automobile accident when the victim's car is struck violently from the rear. Explain why the head of the victim seems to be thrown backward in this situation. Is it really?

ANSWER: When the car is struck from the rear, the force on the car causes the car to move forward. The seat and seatback are attached to the body of the car and move forward with the car. As the seatback moves forward it pushes the occupant's torso forward. The person's head is above the seatback and the head's inertia causes the head to appear to be "thrown backward" as the car moves forward. In reality, the person's head tends to remain at rest while the car plus victim's torso is pushed forward.

Modern cars are equipped with a headrest which, if properly positioned, will push the head forward along with the torso if the car is struck from the rear.

Weight

The force of gravity on an object, i.e., the **weight**, varies slightly from place to place on the surface of the Earth. The variation is so small that the weight of an object located close to the Earth's surface is usually considered to be constant.

The formula for an object's weight ($\vec{\mathbf{w}}$ or $\vec{\mathbf{F}}_g$) near the Earth's surface can be determined as follows: $\Sigma\vec{\mathbf{F}} = m\,\vec{\mathbf{a}}$, but $\Sigma\vec{\mathbf{F}} = \vec{\mathbf{F}}_g = \vec{\mathbf{w}}$ and $\vec{\mathbf{a}} = \vec{\mathbf{g}}$; therefore, $\vec{\mathbf{F}}_g = \vec{\mathbf{w}} = m\,\vec{\mathbf{g}}$

Mass and weight are often confused. Mass depends on an object's inertia and does not vary with location. Weight is the force of gravity acting on an object and varies from place to place. For example, a person standing on the Earth weighs six times as much as he or she would weigh on the Moon; however, his or her mass would remain the same.

EXAMPLE PROBLEM 1. A box of mass 5.0 kg is pulled vertically upward by a force of 68 newtons applied to a rope attached to the box. Determine the a) rate of acceleration of the box and b) vertical velocity of the box after 2.0 s of motion.

Part a. Step 1.	Solution: (Sections 4-4 and 4-7)
Draw an accurate diagram locating the forces acting on the box, then draw a free body diagram.	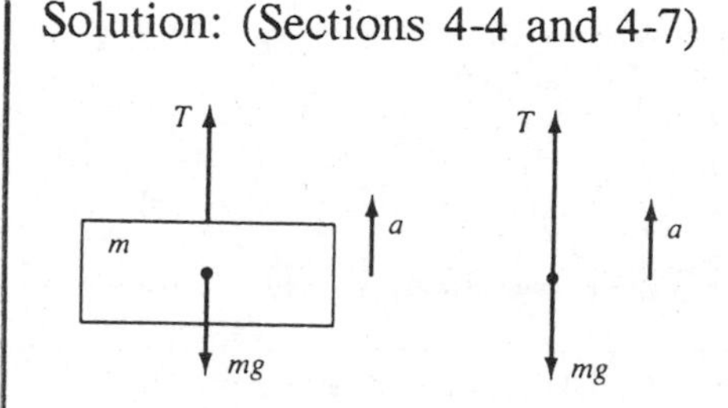

Part a. Step 2.

Determine the net force acting on the box.

The box moves upward; therefore, the tension (T) is greater than the weight, therefore the net force is

$T - mg = 68\ N - (5.0\ kg)(9.8\ m/s^2) = 19\ N$

Part a. Step 3.

Apply Newton's second law of motion and solve for the rate of acceleration.

$\Sigma \vec{F} = m\,\vec{a}$

$68\ N - (5.0\ kg)(9.8\ m/s^2) = (5.0\ kg)\ a$

$19\ N = (5.0\ kg)\ a$

$\vec{a} = 3.8\ m/s^2$

Part b. Step 1.

Determine the velocity the box after 2.0 s.

The acceleration is uniform; therefore, the kinematics equations derived for uniformly accelerated motion can be used to determine the velocity.

$v = at + v_o$

$v = (3.8\ m/s^2)(2.0\ s) + 0\ m/s$

$v = 7.6\ m/s$, upward

EXAMPLE PROBLEM 2 A student of mass 50 kg decides to test Newton's laws of motion by standing on a bathroom scale placed on the floor of an elevator. Assume that the scale reads in newtons. Determine the scale reading when the elevator is a) accelerating upward at $0.50\ m/s^2$, b) traveling upward with a constant speed of 3.0 m/s, and c) traveling upward but decelerating at $1.0\ m/s^2$.

Part a. Step 1.

Draw a diagram locating the forces acting on the student.

Solution: (Sections 4-4, 4-7)

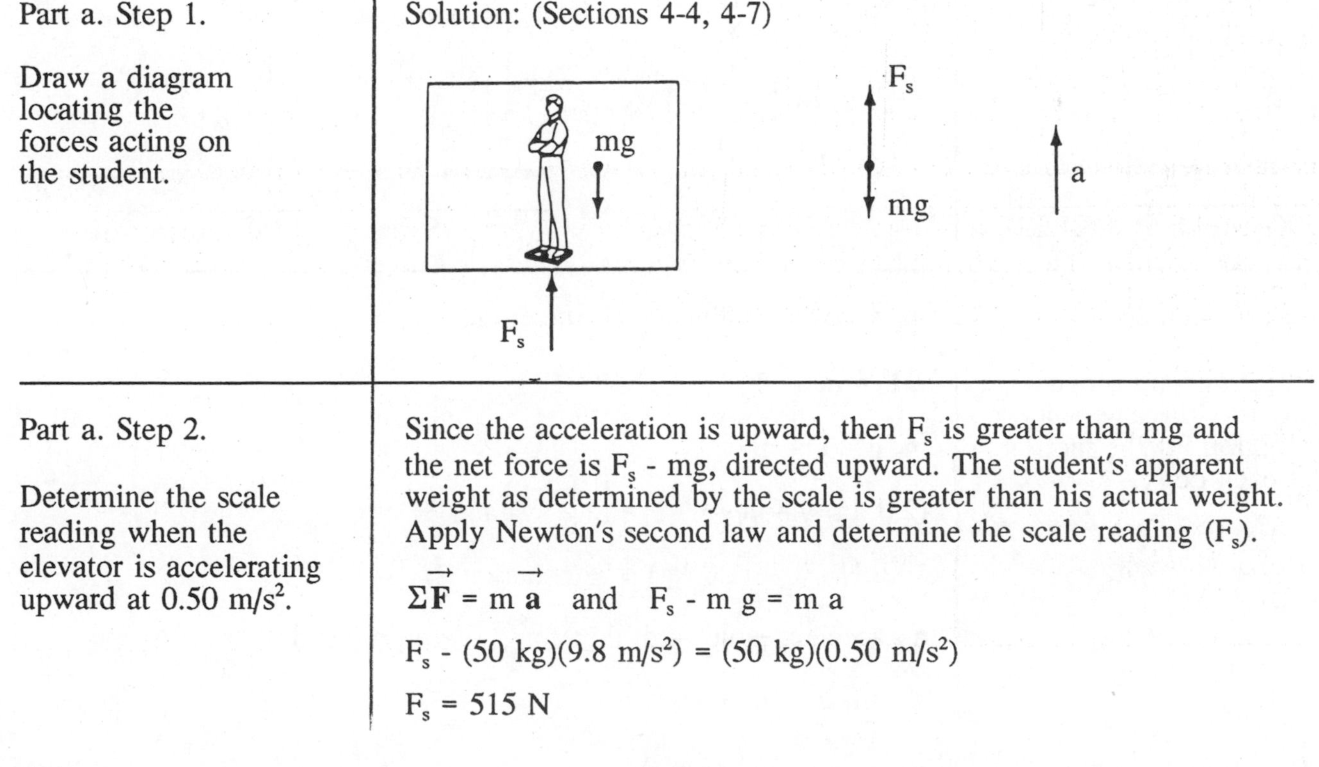

Part a. Step 2.

Determine the scale reading when the elevator is accelerating upward at $0.50\ m/s^2$.

Since the acceleration is upward, then F_s is greater than mg and the net force is F_s - mg, directed upward. The student's apparent weight as determined by the scale is greater than his actual weight. Apply Newton's second law and determine the scale reading (F_s).

$\Sigma \vec{F} = m\,\vec{a}$ and $F_s - m\,g = m\,a$

$F_s - (50\ kg)(9.8\ m/s^2) = (50\ kg)(0.50\ m/s^2)$

$F_s = 515\ N$

Part b. Step 1. Determine the reading on the scale when the elevator is traveling upward at a constant speed of 3.0 m/s.	The velocity is constant; therefore, the rate of acceleration is zero and the net force must equal zero. The student's apparent weight as read by the scale equals his actual weight. $\Sigma \vec{F} = m\,\vec{a}$ and $F_s - mg = m\,a$ but $a = 0\ m/s^2$ $F - (50\ kg)(9.8\ m/s^2) = (50\ kg)(0\ m/s^2)$ $F = 490\ N$
Part c. Step 1. Determine the reading on on the scale when it is traveling upward but decelerating at $1.0\ m/s^2$.	The elevator is decelerating; therefore, the downward force of his weight is greater than the upward force exerted by the scale. Therefore, the student's apparent weight as determined by the scale is less than his actual weight. $\Sigma \vec{F} = m\,\vec{a}$ and $F_s - mg = m\,a$ $F_s - (50\ kg)(9.8\ m/s^2) = (50\ kg)(-\ 1.0\ m/s^2)$ $F_s = 440\ N$

EXAMPLE PROBLEM 3. In a device known as an Atwood machine, a massless, unstretchable rope passes over a frictionless peg. One end of the rope is connected to object $m_1 = 1.0$ kg while the other end is connected to object $m_2 = 2.0$ kg. The system is released from rest and the 2.0 kg object accelerates downward while the 1.0 kg object accelerates upward. Calculate the a) rate of acceleration and b) tension in the rope.

Part a. Step 1.

Draw an accurate diagram locating the forces acting on each object.

Solution: (Section 4-7)

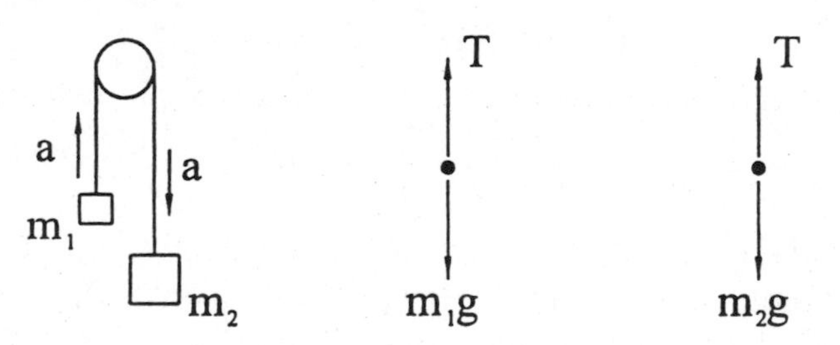

Part a. Step 2.

Use Newton's second law to write a separate equation for the motion of each object.

m_2 accelerates downward; therefore, $m_2g > T$.

$\Sigma \vec{F} = m\,\vec{a}$

$m_2\,g - T = m_2\,a$

$(2.0\ kg)(9.8\ m/s^2) - T = (2.0\ kg)\,a$

$19.6\ N - T = (2.0\ kg)\,a$ (equation 2)

m_1 accelerates upward; therefore, $T > m_1\,g$ and

$T - m_1\,g = m_1\,a$

	$T - (1.0\ kg)\ (9.8\ m/s^2) = (1.0\ kg)\ a$ $T - 9.8\ N = (1.0\ kg)\ a$ (equation 1)
Part a. Step 3. Use algebra to solve for the rate of acceleration of the system.	The above equations have the same two unknowns. To solve for The acceleration, add the two equations together. $19.6\ N - T + T - 9.8\ N = (2.0\ kg + 1.0\ kg)\ a$ $9.8\ N = (3.0\ kg)\ a$ and $a = 3.3\ m/s^2$
Part a. Step 4. Solve for the acceleration using only the external forces acting on the system.	Tension is an internal force in the system. As can be seen in Step 3, the tension algebraically cancels when solving for the acceleration. The external force that causes the motion is the weight of m_2 while the weight of m_1 retards the motion. Applying Newton's second law gives $\Sigma \vec{F} = m \vec{a}$ $m_2 g - m_1 g = (m_2 + m_1)\ a$ $(2.0\ kg)(9.8\ m/s^2) - (1.0\ kg)(9.8\ m/s^2) = (2.0\ kg + 1.0\ kg)\ a$ $19.6\ N - 9.8\ N = (3.0\ kg)\ a$ $a = 3.3\ m/s^2$
Part b. Step 1. Solve for the tension in the rope.	From equation 1: $T - 9.8\ N = (1.0\ kg)\ a$ $T - 9.8\ N = (1.0\ kg)(3.3\ m/s^2)$ $T = 13\ N$

Friction

Friction is a contact force that opposes the relative motion of two surfaces as they slide past each other. The maximum value of the frictional force (F_{fr}) is proportional to the **coefficient of friction** (μ) and also to the **normal force** (F_N) acting on the object.

$$F_{fr} = \mu\ F_N$$

The coefficient of friction is a pure number without physical units and varies with the types of surfaces that are in contact with each other. The force to overcome friction when the object is initially at rest is greater than when the object is moving. When the object is at rest, the **coefficient of static friction** (μ_s) is used to determine the magnitude of the frictional force just before the object starts to move. When the object is moving, the **coefficient of kinetic friction** (μ_k) is used.

The normal force is a contact force that acts perpendicular to the common surface of contact. For example, a book at rest on a horizontal table is acted upon by two forces. The book's weight is downward but is opposed by the normal force which is equal in magnitude but directed upward, perpendicular to the surface of the table.

TEXTBOOK QUESTION 19. A heavy crate rests on the bed of a flatbed truck. When the truck accelerates, the crate remains where it is on the truck, so it too accelerates. What force causes the crate to accelerate.

ANSWER: The net force acting on the crate is due to static friction. In order for the crate to remain at rest relative to the surface of the bed of the truck, the frictional force must be sufficient to cause the crate to accelerate at the same rate as the truck. The maximum value of the frictional force is related to the coefficient of static friction as well as the normal force. If the rate of acceleration of the truck exceeds a certain value, the frictional force will be insufficient to accelerate the crate at the same rate as the truck and the crate will slide toward the back of the truck.

EXAMPLE PROBLEM 4. A hockey puck of mass 0.50 kg traveling at 10 m/s on a horizontal surface slows to 2.0 m/s over a distance of 80 m. Determine the a) magnitude of the frictional force acting on the puck and b) coefficient of kinetic friction between the puck and the surface.

Part a. Step 1. Draw an accurate diagram locating the forces acting on the puck, then draw a free body diagram.	Solution: (Sections 4-4 and 4-7) v_0 F_{fr} F_N a mg
Part a. Step 2. Complete a data table.	m = 0.50 kg, x = 80 m v_o = 10 m/s, x_o = 0 m, v = 2.0 m/s
Part a. Step 3. Use the kinematics equations to determine the acceleration.	$v^2 = v_o^2 + 2\,a\,(x - x_o)$ $(2.0\text{ m/s})^2 = (10\text{ m/s})^2 + 2\,a(80\text{ m} - 0)$ $a = -\,0.60\text{ m/s}^2$
Part a. Step 4. Use Newton's second law to determine the magnitude of the frictional force.	$\Sigma\vec{\mathbf{F}} = m\,\vec{\mathbf{a}}$ $F_{fr} = (0.50\text{ kg})(-\,0.60\text{ m/s}^2)$ $F_{fr} = -\,0.30$ newtons Note: the frictional force is negative because it is in the direction opposite from the direction of motion.

Part b. Step 1.

Determine the magnitude and direction of the normal force.

Since the puck is on a level surface, the normal force is equal in magnitude but in the opposite direction from the puck's weight.

$F_N = mg = (0.50\text{ kg})(9.8\text{ m/s}^2)$

$F_N = 4.9\text{ N}$

Part b. Step 2.

Determine the coefficient of kinetic friction between the puck and the ice.

$F_{fr} = \mu_k F_N = \mu_k mg$

$0.30\text{ N} = \mu_k (0.50\text{ kg})(9.8\text{ m/s}^2)$

$\mu_k = 0.061$

EXAMPLE PROBLEM 5. As shown in the diagram, a 10.0-kg box on a level surface is attached by a weightless, unstretchable rope to a 7.00-kg box that rests on a 30.0° incline. The coefficient of kinetic friction between each box and the surface is 0.100. The system is released from rest and both objects accelerate toward the right. Determine the a) rate of acceleration of the system and b) tension in the rope that connects the boxes.

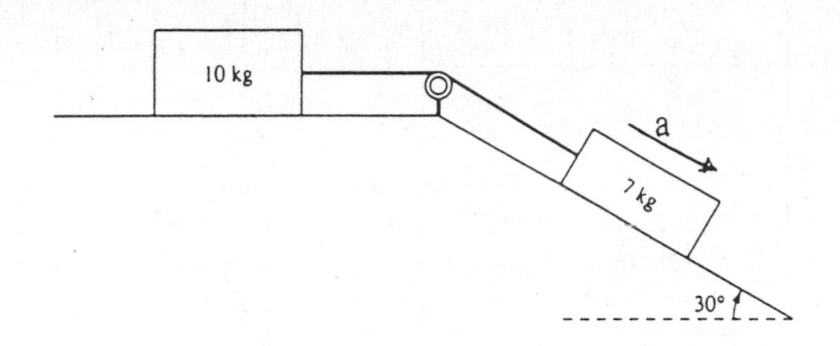

Part a. Step 1.

Draw a free body diagram locating the forces acting on each object. Note: these forces include tension in the rope, weight, normal force, and frictional force acting on each.

Solution: (Sections 4-4 and 4-7)

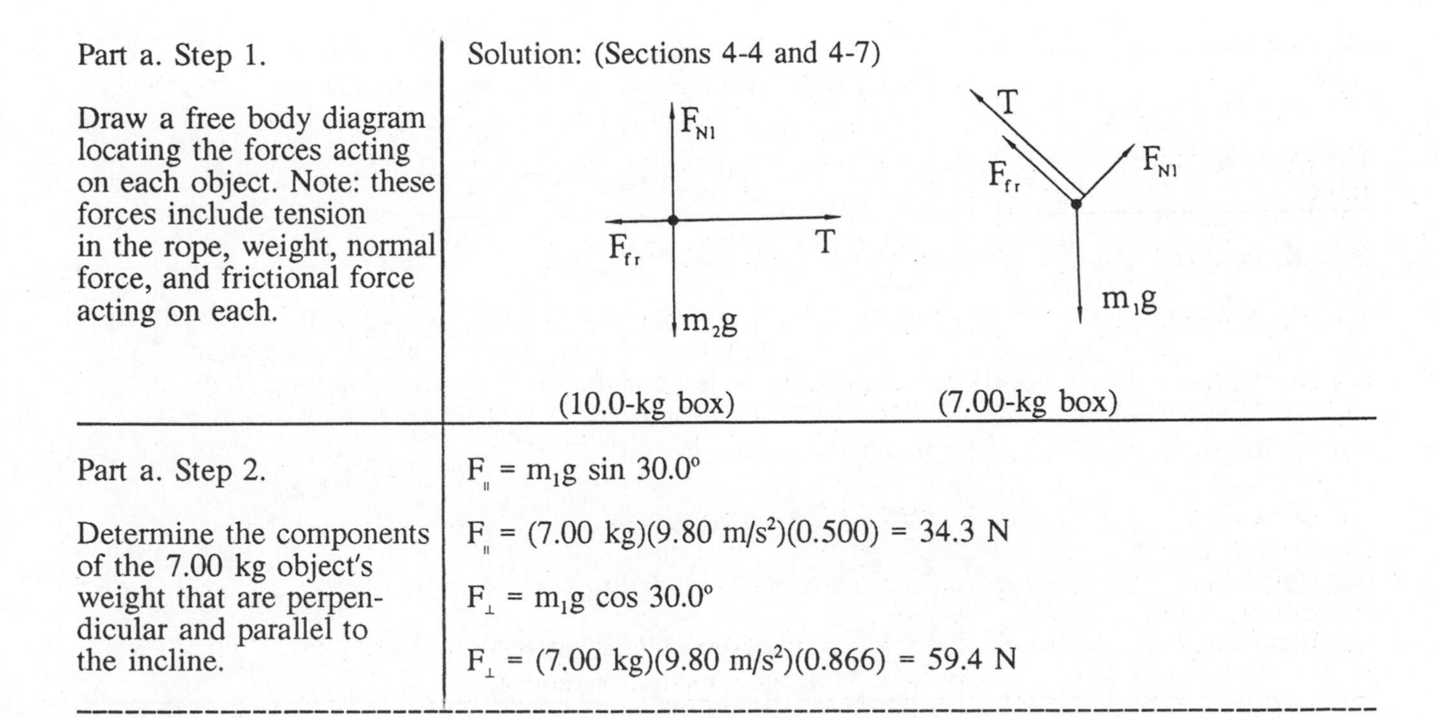

(10.0-kg box) (7.00-kg box)

Part a. Step 2.

Determine the components of the 7.00 kg object's weight that are perpendicular and parallel to the incline.

$F_{\parallel} = m_1 g \sin 30.0°$

$F_{\parallel} = (7.00\text{ kg})(9.80\text{ m/s}^2)(0.500) = 34.3\text{ N}$

$F_{\perp} = m_1 g \cos 30.0°$

$F_{\perp} = (7.00\text{ kg})(9.80\text{ m/s}^2)(0.866) = 59.4\text{ N}$

Part a. Step 3.

Determine the magnitude of the frictional force acting on the box.

The frictional force is given by $F_{fr} = \mu_k F_N$, where F_N is equal in magnitude but opposite in direction to $F_{\perp}$.

For the box on the horizontal surface, the force of friction

$F_{fr} = \mu_k F_N = \mu_k m_2 g \cos 0^\circ$

$F_{fr} = (0.100)(10.0 \text{ kg})(9.80 \text{ m/s}^2)(1.00) = 9.80 \text{ N}$

For the box on the 30.0° incline, the frictional force

$F_{fr} = \mu_k m_1 g \cos 30.0^\circ = (0.100)(59.4 \text{ N})$

$F_{fr} = 5.94 \text{ N}$

Part a. Step 4.

Use Newton's second law to write a separate equation for the motion of each object.

The motion of object 1 is parallel to the incline, i.e., $F_{\parallel}$ causes the motion while F_{fr} and the tension in the string retard the motion.

object 1:
(7.00 kg box)

$F_{\parallel} - F_{fr} - T = m_1 a$

$34.3 \text{ N} - 5.94 \text{ N} - T = (7.00 \text{ kg}) a$

$28.4 \text{ N} - T = (7.00 \text{ kg}) a$

The motion of the 10.0 kg box, i.e., object 2, is along the level surface. The tension in rope causes object 2 to accelerate while friction retards its motion.

object 2:
(10.0 kg box)

$T - F_{fr} = m_2 a$

$T - 9.80 \text{ N} = (10.0 \text{ kg}) a$

Part a. Step 5.

Use algebra to solve for the rate of acceleration of the system.

The above equations have the same two unknowns. To solve for the acceleration, add the two equations together.

$28.4 \text{ N} - T + T - 9.80 \text{ N} = (7.00 \text{ kg} + 10.0 \text{ kg}) a$

$18.6 \text{ N} = (17.0 \text{ kg}) a$ and $a = 1.09 \text{ m/s}^2$

Part a. Step 6.

Solve for the acceleration considering only the external forces acting on the system.

Tension is an internal force and the problem can be solved by considering only the external forces acting on the system. The forces acting on the system are $F_{\parallel}$ and the retarding force of by friction acting on each box. Apply the second law and solve for the acceleration.

Net $F = m a$

$F_{\parallel} - F_{fr1} - F_{fr2} = (m_1 + m_2) a$

$34.4 \text{ N} - 5.97 \text{ N} - 9.80 \text{ N} = (7.0 \text{ kg} + 10.0 \text{ kg}) a$

$18.5 \text{ N} = (17.0 \text{ kg}) a$ and $a = 1.09 \text{ m/s}^2$

Part b. Step 1.	$T - 9.80\ N = (10.0\ kg)(1.09\ m/s^2)$
Solve for the tension in the rope.	$T = 20.8\ N$

PROBLEM SOLVING SKILLS

For problems related to Newton's second law of motion:

1. Draw an accurate diagram locating each of the forces acting on the object or system of objects.
2. Draw a free body diagram locating the forces acting on the object(s) in question.
3. If the object is on an incline, the weight is replaced with components acting parallel and perpendicular to the incline.
4. If a frictional force is involved, the magnitude of the force is related to the coefficient of friction and the normal force.
5. Determine the magnitude of the net force acting on the object.
6. Use Newton's second law to write an equation for the motion of each object in the system. Solve for the rate of acceleration of each object in the system.
7. If the problem involves uniform acceleration, then the kinematics equations developed for uniformly accelerated motion can be used to determine the velocity, displacement, time, etc.

SOLUTIONS TO SELECTED TEXTBOOK PROBLEMS

TEXTBOOK PROBLEM 10. How much tension must a rope withstand if it is used to accelerate a 1200 kg car vertically upward at 0.80 m/s^2?

Part a. Step 1.

Draw a free body diagram locating the forces acting on the car.

Solution: (Sections 4-4 and 4-6)

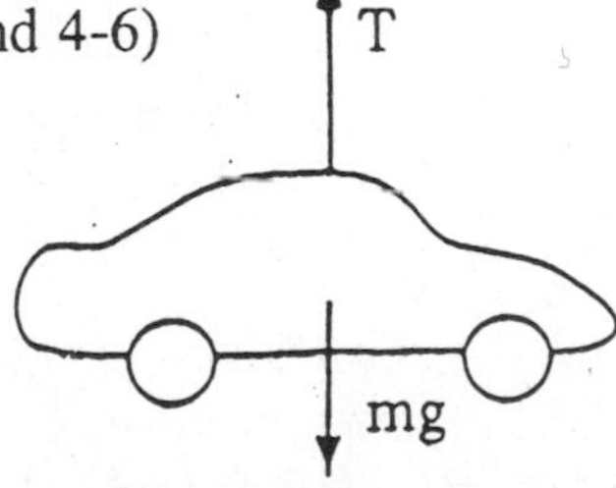

Part a. Step 2.

Apply Newton's second law and solve for the tension.

The car accelerates upward; therefore, the tension (T) is greater than the weight, therefore the $\Sigma F = T - mg$.

$\Sigma F = ma$

$T - (1200\ kg)(9.8\ m/s^2) = (1200\ kg)(0.80\ m/s^2)$

$T - 11760\ N = 960\ N$

$T = 1.3 \times 10^4\ N$

TEXTBOOK PROBLEM 37. A force of 48.0 N is required to start a 5.0-kg box moving across a horizontal concrete floor. (a) What is the coefficient of static friction between the box and the floor? (b) If the 48.0 N force continues, the box accelerates at 0.70 m/s^2. What is the coefficient of kinetic friction?

Part a. Step 1.

Draw a free body diagram locating the forces acting on the box.

Solution: (Section 4-8)

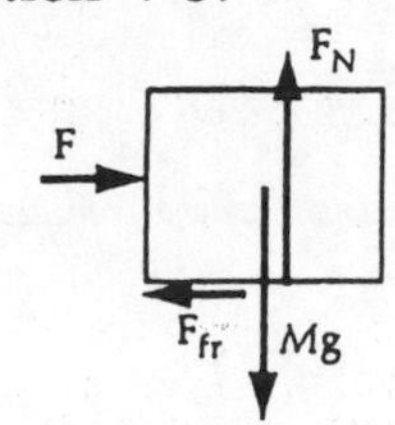

Part a. Step 2.

Use Newton's second law to solve for the coefficient of static friction.

Note: the force of static friction $F_{fr} \leq \mu_s F_N$. When the box is just about to move $F_{fr} = \mu_s F_N$ where $F_N = m g$. At that point the rate of acceleration is zero.

$\Sigma \vec{\mathbf{F}} = m \vec{\mathbf{a}}$

$F - F_{fr} = ma$ but $a = 0 \text{ m/s}^2$ and $F_{fr} = \mu_s m_1 g$

$48.0 \text{ N} - \mu_s (5.0 \text{ kg})(9.8 \text{ m/s}^2) = 0 \text{ m/s}^2$

$- \mu_s = (- 48.0 \text{ N})/(49 \text{ N})$

$\mu_s = 0.98$

Part b. Step 1.

Use Newton's second law to solve for the coefficient of kinetic friction (μ_k).

When the box begins to move, kinetic friction opposes the motion.

$\Sigma \vec{\mathbf{F}} = m \vec{\mathbf{a}}$

$F - F_{fr} = m a$ where $F_{fr} = \mu_k m_1 g$

$48.0 \text{ N} - \mu_k (5.0 \text{ kg})(9.8 \text{ m/s}^2) = (5.0 \text{ kg})(0.70 \text{ m/s}^2)$

$48.0 \text{ N} - \mu_k (49 \text{ N}) = 3.5 \text{ N}$

$- \mu_k = (- 48.0 \text{ N} + 3.5 \text{ N})/(49 \text{ N})$

$\mu_k = 0.91$

TEXTBOOK PROBLEM 52. A carton shown in Fig. 4-55 lies on a plane tilted at an angle $\theta = 22.0°$ to the horizontal, with $\mu_k = 0.12$. (a) Determine the rate of acceleration of the carton as it slides down the plane. (b) If the carton starts from rest 9.30 m up the plane from the base, what will be the carton's speed when it reaches the bottom of the incline?

Part a. Step 1.

Draw a free body diagram locating all of the forces acting on the carton.

Solution: (Section 4-8)

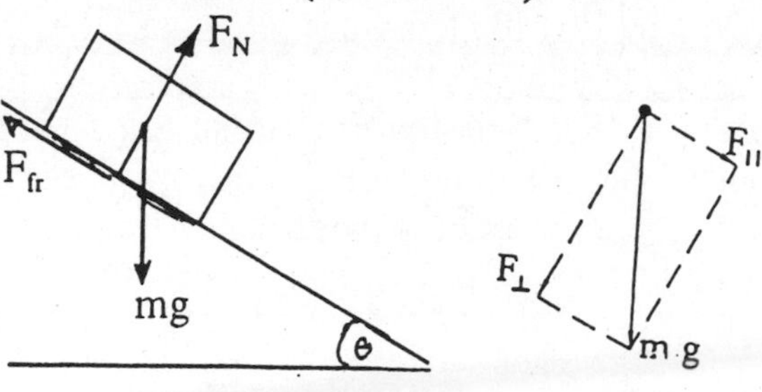

Part a. Step 2. Determine the components of the object's weight perpendicular and parallel to the incline.	$F_{\parallel}/mg = \sin 22.0^\circ$ and $F_{\parallel} = mg \sin 22.0^\circ = 0.375$ mg $F_{\perp}/mg = \cos 22.0^\circ$ and $F_{\perp} = mg \cos 22.0^\circ = 0.927$ mg
Part a. Step 3. Determine the magnitude and direction of the carton's acceleration.	There is no motion perpendicular to the incline; the magnitude of the normal force equals the perpendicular component of the weight. The frictional force is related to the perpendicular component of the weight by the formula $F_{fr} = \mu_k F_N$, where F_N is equal in magnitude but opposite in direction to $F_{\perp}$. The component of the weight parallel to the incline causes the motion. Since $F_{\parallel}$ is greater than F_{fr}, the $\Sigma F = F_{\parallel} - F_{fr}$. $\Sigma \vec{\mathbf{F}} = m\,\vec{\mathbf{a}}$ $F_{\parallel} - F_{fr} = m\,a$ where $F_{fr} = \mu_k F_N = \mu_k mg \cos\theta$ $(0.375)\,mg - (0.12)(0.927)\,mg = m\,a$ Note that the mass of the object appears in each term. Dividing each term by m gives $(9.8\ m/s^2)(0.375) - (0.12)(9.8\ m/s^2)(0.927) = a$ $3.68\ m/s^2 - 1.09\ m/s^2 = a$ $a = 2.6\ m/s^2$
Part b. Step 1. Determine the object's speed at the bottom of the incline. Note: the initial speed at the top $v_o \approx 0$.	$2\,a\,(x - x_o) = v^2 - v_o^2$ $2\,(2.6\ m/s^2)(9.30\ m) = v^2 - (0\ m/s)^2$ $v^2 = 48.4\ m^2/s^2$ $v = 7.0\ m/s$

TEXTBOOK PROBLEM 64. (a) Suppose the coefficient of kinetic friction between the m_1 and the plane shown in Fig. 4-57 is $\mu_k = 0.150$ and that $m_1 = m_2 = 2.70$ kg. As m_2 moves down, determine the magnitude of the acceleration of m_1 and m_2 given $\theta = 25.0^\circ$. (b) What smallest value of μ_k will keep this system from accelerating?

Part a. Step 1.

Draw a free body diagram locating the forces acting on each object. Note: these forces include tension in the rope, the weight, the normal force, and frictional force acting on each object.

Solution: (Section 4-8)

T, m_2g — object m_2

T, F_N, F_{fr}, θ, m_1g — object m_1

Part a. Step 2.

Determine the components of m_1's weight that are perpendicular and parallel to the incline.

$F_{\parallel}/m_1g = \sin 25.0°$

$F_{\parallel} = m_1g \sin 25.0° = (2.70 \text{ kg})(9.80 \text{ m/s}^2)(0.423) = 11.2 \text{ N}$

$F_{\perp}/m_1g = \cos 25.0°$

$F_{\perp} = m_1g \cos 25.0° = (2.70 \text{ kg})(9.80 \text{ m/s}^2)(0.906) = 24.0 \text{ N}$

Part a. Step 3.

Determine the magnitude of the frictional force (F_{fr}) acting on m_1.

For the box on the 25.0° incline, $F_{fr} = \mu_k F_N$, where F_N is equal in magnitude but opposite in direction to $F_{\perp}$.

$F_{fr} = \mu_k m_1g \cos 25.0° = (0.150)(24.0 \text{ N}) = 3.60 \text{ N}$

Part a. Step 4.

Use Newton's second law to write a separate equation for the motion of each object.

The motion of object 1 is parallel to the incline, i.e., the tension (T) in the string causes the motion while $F_{\parallel}$ and F_{fr} retard the motion.

object 1: $T - F_{\parallel} - F_{fr} = m_1 a$

$T - 11.2 \text{ N} - 3.60 \text{ N} = (2.70 \text{ kg})\, a$

(equation 1) $T - 14.8 \text{ N} = (2.70 \text{ kg})\, a$

The motion of m_2 is directed downward. The weight of m_2 causes the motion while the tension in the string opposes the motion.

object 2: $m_2 g - T = m_2 a$

$(2.70 \text{ kg})(9.80 \text{ m/s}^2) - T = (2.70 \text{ kg})\, a$

(equation 2) $26.5 \text{ N} - T = (2.70 \text{ kg})\, a$

Part a. Step 5.

Use algebra to solve for the rate of acceleration of the system.

The above equations have the same two unknowns. To solve for the acceleration, add the two equations together.

$26.5\ \text{N} - T + T - 14.8\ \text{N} = (2.70\ \text{kg})\ a + (2.70\ \text{kg})\ a$

$11.7\ \text{N} = (5.40\ \text{kg})\ a$ and $a = 2.17\ \text{m/s}^2$

Part b. Step 1.

Determine the smallest value of μ_k which will keep this system from accelerating.

From part a. step 4. object 1: $T - F_{\parallel} - F_{fr} = m_1\ a$

object 2: $m_2\ g - T = m_2\ a$

Adding the two equations together gives

$m_2\ g - F_{\parallel} - F_{fr} = m_1\ a + m_2\ a$ but $a = 0$

$m_2 g - m_1 g\ \sin 25.0° - \mu_s\ m_1 g \cos 25.0° = 0$ and $m_2 g = m_1 g$;

$1 - \sin 25.0° - \mu_k \cos 25.0° = 0$

$1 - 0.423 - \mu_k\ (0.906) = 0$ and solving for μ_k gives

$\mu_k = (0.577)/(0.906) = 0.64$

TEXTBOOK PROBLEM 69. A 1150 kg car pulls a 450 kg trailer. The car exerts a horizontal force of 3.8×10^3 N against the ground in order to accelerate. What force does the car exert on the trailer. Assume an effective friction coefficient of 0.15 for the trailer.

Part a. Step 1.

Draw a free body diagram locating the forces acting on each object. Note: these forces include tension in the connection between the car and the trailer, weight, normal force, and frictional force acting on the trailer.

Solution: (Section 4-8)

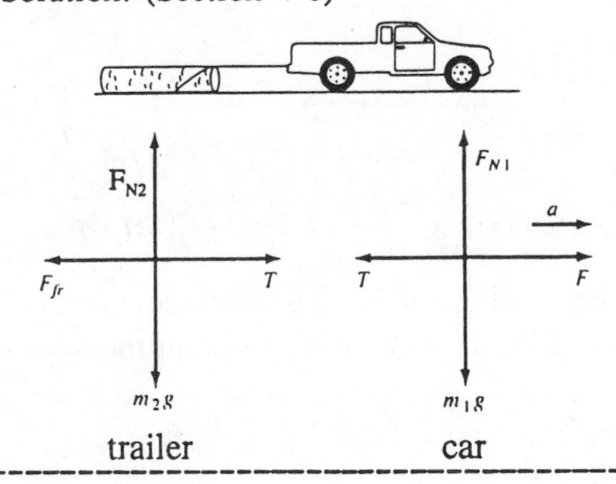

Part a. Step 2.

Use Newton's second law to write a separate equation for the motion of each object.

$\Sigma F = m\ a$

$F_{car} - T = m_{car}\ a$ (car)

$3.8 \times 10^3\ \text{N} - T = (1150\ \text{kg})\ a$

$T - F_{fr} = m_{trailer}\ a$ (trailer)

	$T - \mu_k\, m_{trailer}\, g \cos 0° = m_{trailer}\, a$
	$T - (0.15)(450\text{ kg})(9.80\text{ m/s}^2)(1.00) = (450\text{ kg})\, a$
	$T - 660\text{ N} = (450\text{ kg})\, a$
Part a. Step 3. Use algebra to solve for the rate of acceleration of the system.	The above equations have the same two unknowns. To solve for the acceleration, add the two equations together. $3.8 \times 10^3\text{ N} - T + T - 660\text{ N} = (1150\text{ kg} + 450\text{ kg})\, a$ $3.1 \times 10^3\text{ N} = (1600\text{ kg})\, a$ and $a = 2.0\text{ m/s}^2$
Part b. Step 1. Solve for the tension in the connection between the car and the trailer.	$T - 660\text{ N} = (450\text{ kg})\, a$ $T - 660\text{ N} = (450\text{ kg})(2.0\text{ m/s}^2)$ $T = 1.54 \times 10^3\text{ N}$

TEXTBOOK PROBLEM 76. A 28.0-kg block is connected to an empty 1.35-kg bucket by a cord running over a frictionless pulley (Fig. 4-59). The coefficient of static friction between the table and the block is 0.450 and the coefficient of kinetic friction between the table and the block is 0.320. Sand is gradually added to the bucket until the system just begins to move. (a) Calculate the mass of sand added to the bucket. (b) Calculate the acceleration of the system.

Part a. Step 1.

Draw a free body diagram locating the forces acting on each object. Note: let m_2 represent the mass of the bucket + sand while m_1 represents the mass of the block.

Solution: (Section 4-7)

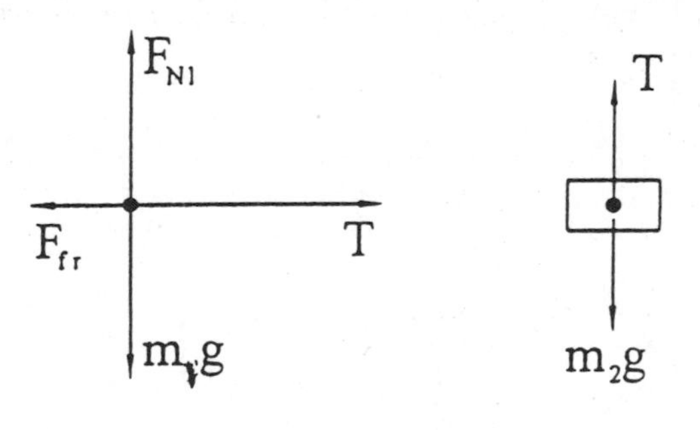

Part a. Step 2.

Use Newton's second law to write a separate equation for the motion for each object.

Once enough sand is added, the bucket + sand accelerate downward, therefore, $m_2 g > T$

Net $F = m\, a$

$m_2\, g - T = m_2\, a$ (equation 1)

The normal force F_N that the table exerts upward balances the weight of object 1. However, a frictional force F_{fr} opposes the motion of m_1. Since m_1 accelerates horizontally across the table, $T > F_{fr}$ and

	$T - F_{fr} = m_1\, a$ where $F_{fr} = \mu_s\, F_N = \mu_s\, m_1\, g$ (equation 2)
Part a. Step 3. Use algebra to solve for the mass m_2 required to just cause the system to move.	At the point where motion is *about* to begin, the force of static friction will be a maximum but the rate of acceleration equals zero. The above equations have the same two unknowns. To solve for m_2, add the two equations together. $m_2\, g - T + T - \mu_s\, m_1\, g = (m_2 + m_1)\, a$ $m_2\, g - \mu_s\, m_1\, g = (m_2 + m_1)\, a$ (equation 3) but $a = 0\ m/s^2$ $m_2\, (9.8\ m/s^2) - (0.450)(28.0\ kg)(9.8\ m/s^2) = 0$ $m_2 = (123\ N)/(9.8\ m/s^2) = 12.6\ kg$ m_2 equals the mass of the bucket + sand. Since the mass of the bucket is 1.35 kg, the mass of sand is 11.3 kg.
Part b. Step 1. Use equation 3 from Part a. Step 3 to solve for the rate of acceleration.	Once the system begins to move, the motion is opposed by kinetic friction instead of static friction; i.e., $F_{fr} = \mu_k\, F_N = \mu_k\, m_1\, g$. $m_2\, g - \mu_k\, m_1\, g = (m_2 + m_1)\, a$ $(12.6\ kg)(9.8\ m/s^2) - (0.320)(28.0\ kg)(9.8\ m/s^2) = (12.6\ kg + 28.0\ kg)\, a$ $123\ N - 87.8\ N = (40.6\ kg)\, a$ $a = 0.879\ m/s^2$

CHAPTER 5

CIRCULAR MOTION; GRAVITATION

OBJECTIVES

After studying the material of this chapter, the student should be able to:

- calculate the centripetal acceleration of a point mass in uniform circular motion given the radius of the circle and either the linear speed or the period of the motion.
- identify the force that is the cause of the centripetal acceleration and determine the direction of the acceleration vector.
- use Newton's laws of motion and the concept of centripetal acceleration to solve word problems.
- distinguish between centripetal acceleration and tangential acceleration.
- state the relationship between the period of the motion and the frequency of rotation and express this relationship using a mathematical equation.
- write the equation for Newton's universal law of gravitation and explain the meaning of each symbol in the equation.
- determine the magnitude and direction of the gravitational field strength (g) at a distance r from a body of mass m.
- use Newton's second law of motion, the universal law of gravitation, and the concept of centripetal acceleration to solve problems involving the orbital motion of satellites.
- explain the "apparent" weightlessness of an astronaut in orbit.
- state from memory Kepler's laws of planetary motion.
- use Kepler's third law to solve word problems involving planetary motion.
- use Newton's second law of motion, the universal law of gravitation, and the concept of centripetal acceleration to derive Kepler's third law.
- solve word problems related to Kepler's third law.
- identify the four forces that exist in nature.

KEY TERMS AND PHRASES

uniform circular motion occurs when an object travels in a circle at constant speed.

tangential velocity refers to the direction of the velocity vector of an object as it travels in a curved path. The velocity vector is tangent to the path at each point in the curve.

tangential acceleration refers to the change in speed per unit time of an object as it travels in a curved path. The tangential acceleration is directed tangent to the path at each point.

centripetal acceleration or radial acceleration is the inward radial acceleration experienced by an object traveling in a circle. The direction of the acceleration vector is perpendicular to the direction of the velocity vector.

period (T) of the circular motion is the time required for an object to complete one revolution.

frequency of rotation (f) is the number of repetitions of the motion per unit time. The frequency is inversely related to the period of the motion.

centripetal force is the term used for the force acting toward the center of the circle. The centripetal force is not some new kind of force. The term merely means that the force is directed toward the circle's center. The force must be applied by some external agent.

Newton's universal law of gravitation is the force of attraction between any two objects due to their mass. The law states that the magnitude of the force between two particles is directly proportional to the product of their masses and inversely proportional to the square of the distance between their centers.

Kepler's first law states that the path of each planet about the Sun is an ellipse with the Sun at one of the focal points of the ellipse.

Kepler's second law states that each planet moves such that an imaginary line drawn from the Sun to the planet sweeps out equal areas in equal periods of time.

Kepler's third law states that the ratio of the squares of the periods of any two planets revolving about the Sun is equal to the ratio of the cubes of their average distances from the Sun.

SUMMARY OF MATHEMATICAL FORMULAS

centripetal or radial acceleration	$a_R = v^2/r$	Centripetal or radial acceleration (a_R) as related to tangential velocity (v) and the radius of the circle (r).
frequency and period of rotation	$f = 1/T$ $T = 1/f$	Frequency of rotation (f) is inversely related to the period (T) of the motion.
tangential velocity	$v = 2\pi r/T$ $v = 2\pi r f$	Tangential velocity (v) as related to the radius of the circle and the period (T) or frequency (f) of the motion.
centripetal or radial acceleration	$a_R = 4\pi^2 r/T^2$ $a_R = 4\pi^2 r f^2$	Centripetal or radial acceleration (a_R) as related to the radius of the circle and the period (T) or frequency (f) of the motion.
centripetal force	$\Sigma F_R = m v^2/r$	Centripetal force (F_R) as related to the object's mass (m), speed (v), and the radius of the circle.

Newton's universal law of gravitation	$F = G\, m_1\, m_2/r^2$	Gravitational force between two particles of mass m_1 and m_2 separated by a radial distance (r). G = universal constant = 6.67×10^{-11} N m^2/kg^2.
Kepler's third law	T^2/r^3 = constant or $T_1^2/T_2^2 = r_1^3/r_2^3$	Period of the motion (T) squared of a planet traveling about the Sun divided by the average radius (r) of its orbit cubed is a constant for all planets.

CONCEPT SUMMARY

Kinematics of Uniform Circular Motion

An object traveling in a circle at constant speed is exhibiting **uniform circular motion.** Although the object's speed is not changing, its direction of motion is and therefore its velocity is constantly changing. Note: recall that speed is a scalar quantity having only magnitude while velocity is a vector quantity having both magnitude and direction.

Since its velocity is changing, the object is undergoing accelerated motion. This acceleration is called **centripetal acceleration.** As shown in the diagram, the velocity vector at every point is directed tangent to the circle while the acceleration vector is directed toward the center of the circle. The direction of the acceleration vector is perpendicular to the direction of the velocity vector.

The magnitude of the centripetal acceleration (a_R) is directly proportional to the square of the speed and inversely proportional to the radius of the circle in which the object is traveling.

$$a_R = v^2/r$$

An object moving in uniform circular motion travels a distance equal to the circumference of the circle ($2\pi r$) in a time interval called the **period** (T) of the motion. The period of the motion is the time required for the motion to repeat. The equation for the speed is as follows:

$$v = 2\pi r/T$$

By substituting the above expression for v into the centripetal acceleration equation, it can be shown that the formula for the centripetal acceleration can be written as follows:

$$a_R = (2\pi r/T)^2/r = 4\,\pi^2\, r/T^2$$

TEXTBOOK QUESTION 2. Will the acceleration of a car be the same if it travels around a sharp curve at a constant 60 km/h as when it travels around a gentle curve at the same speed? Explain.

ANSWER: The acceleration is *not* the same. The centripetal acceleration is directly proportional to the square of the speed and inversely proportional to the radius of the curve. The speed is held constant at 60 km/h but the radius of the sharp curve is less than the gentle curve. Therefore, the *smaller* the radius, the *greater* the centripetal acceleration.

Frequency of Rotation

The **frequency of rotation** (f) is the number of repetitions of the motion per unit time. For example, a stereo turntable rotating at 45 rpm has a frequency of rotation of 45 revolutions per minute.

The period of the motion is inversely related to the frequency of rotation ($T = 1/f$ or $f = 1/T$). If the frequency of rotation is 10 revolutions per second, then the period of the motion is 1/10 second or 0.10 seconds.

Since $f = 1/T$, the equation for the speed $v = 2\pi r/T$ can be written as in terms of the frequency as

$v = 2\pi\, r\, f$

Also, the equation for centripetal acceleration $a_R = 4\,\pi^2\, r/T^2$ can be written as $a_c = 4\,\pi^2\, r\, f^2$.

Dynamics of Uniform Circular Motion

An object traveling in a circle must have a net force acting on it to keep it in the circle. The magnitude of the object's acceleration can be determined by using Newton's second law, i.e., $a = \text{net } F/m$. For an object traveling in a circle, the acceleration is the centripetal acceleration. Since the centripetal acceleration is directed toward the center of the circle, the net force, called the **centripetal force,** must also be directed toward the center of the circle. As stated in the textbook, "This force is sometimes called a centripetal force. But be aware that 'centripetal force' does not indicate some new kind of force. The term merely means that the force is directed toward the circle's center. The force must be applied by some object."

In solving problems involving uniform circular motion, it is usually necessary to determine the nature of the force which causes the object to travel in the circular path. For example, when a car rounds an unbanked curve, the centripetal force is related to friction. Gravity provides the centripetal force for a satellite in orbit, while a ball on the end of a string is held in a circular path by the tension in the string.

TEXTBOOK QUESTION 1. Sometimes people say that water is removed from clothes in a spin dryer by centrifugal force throwing the water outward. What is wrong with this statement.

ANSWER: If the wall of the spinning cylinder was solid, then the inward force that the cylinder exerts on the water, i.e., centripetal force, would not allow the water to escape. However, the cylinder has numerous small holes in it and there is no centripetal force acting on the water at these holes. The inertia of the water in the wet clothes tends to cause the water to travel through these small holes. The water actually travels tangent to the circle in which the clothes are spinning and out through the holes. Centrifugal force has nothing to do with removing the water.

EXAMPLE PROBLEM 1. A 4.00 kg box is placed on the floor at the edge of a merry-go-round of radius 3.00 m. The coefficient of static friction between the box and the floor is 0.200. The merry-go-round accelerates from rest and eventually the box slides off the edge. Determine the speed at which this occurs.

Part a. Step 1. Draw an accurate diagram locating all of the forces acting on the object.	Solution: (Section 5-2) F_N F_{fr} mg
Part a. Step 2. Complete a data table.	$m = 4.00$ kg $\mu_s = 0.200$ $r = 3.00$ m $v = ?$
Part a. Step 3. Determine the net force on the object.	The box is held in position by a frictional force that provides the centripetal acceleration. At the point where the box tends to slide off the circle, the force required to hold the box in position "just" exceeds the maximum possible value of the frictional force. The maximum value of the frictional force can now be determined. $F_{fr} = \mu_s F_N$ but $F_N = mg$ $F_{fr} = \mu_s mg = (0.200)(4.00 \text{ kg})(9.80 \text{ m/s}^2)$ $F_{fr} = 7.84$ N and net $F = 7.84$ N
Part a. Step 4. Use Newton's second law and the formula for centripetal acceleration to determine the object's velocity.	$\Sigma \vec{F} = m \vec{a}$ where net $\vec{F} = \vec{F}_{fr}$ and $a = v^2/r$. $7.84 \text{ N} = (4.00 \text{ kg}) v^2/(3.00 \text{ m})$ $v^2 = 5.88 \text{ m}^2/\text{s}^2$ and $v = 2.42$ m/s Note: it can be shown that the mass of the box does not affect the maximum velocity. $F_{fr} = m v^2/r$ but $F_{fr} = \mu_s F_N = \mu_s mg$ $\mu_s mg = m v^2/r$ The mass cancels from both sides of the equation. $\mu_s g = v^2/r$ $v = (\mu_s g r)^{1/2} = [(0.200)(9.80 \text{ m/s}^2)(3.00 \text{ m})]^{1/2}$ and $v = 2.42$ m/s

EXAMPLE PROBLEM 2. A car travels over the crest of a hill at 10.0 m/s. The radius of curvature at the crest is 12.0 m. a) Determine the force exerted by the car seat on a 60.0 kg passenger. b) Determine the minimum speed required for the passenger to feel momentarily "weightless."

Part a. Step 1. Draw a free-body diagram locating the forces acting on the passenger.	Solution: (Section 5-2) 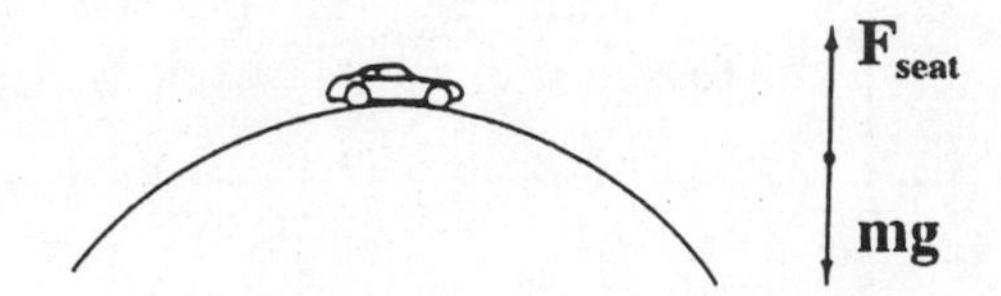
Part a. Step 2. Derive a formula for the net force acting on the person. In which direction is this force acting?	At the crest of the hill the car is traveling in a circle at a constant speed. Since it is traveling in a circle there is a centripetal acceleration directed into the center of the circle. This centripetal acceleration is produced by a net inward force equal to the difference between the passenger's weight, which acts downward, and the upward force exerted by the seat. $\Sigma F = \text{weight} - F_{seat} = mg - F_{seat}$
Part a. Step 3. Apply Newton's second law and the formula for centripetal acceleration and determine the magnitude of the force exerted by the seat.	Since the direction of the net force is downward, let the downward direction be positive. $\Sigma F = m\ a$ $\text{weight} - F_{seat} = m\ a$ but $a = v^2/r$ $(60.0\ \text{kg})(9.80\ \text{m/s}^2) - F_{seat} = (60.0\ \text{kg})[(10.0\ \text{m/s})^2/(12.0\ \text{m})]$ $588\ \text{N} - F_{seat} = 500\ \text{N}$ $F_{seat} = 588\ \text{N} - 500\ \text{N} = 88.0\ \text{N}$ As the car goes over the crest of the hill, the passenger feels less upward force from the seat.
Part b. Step 1. Determine the minimum speed required for the passenger to feel momentarily "weightless."	At this "minimum speed" the seat exerts no upward force on passenger, i.e., $F_{seat} = 0$. As a result the passenger experiences a feeling of momentary "weightlessness." Since the direction of the net force is downward, let the downward direction be positive. $\Sigma F = m\ a$ $\text{weight} - F_{seat} = m\ a$ but $F_{seat} = 0$ and $a = v^2/r$ $(60.0\ \text{kg})(9.80\ \text{m/s}^2) - 0\ \text{N} = (60.0\ \text{kg})\ v^2/(12.0\ \text{m})$ $v = [(9.80\ \text{m/s}^2)(12.0\ \text{m})]^{1/2} = 10.8\ \text{m/s}$, or approximately 24 mph

EXAMPLE PROBLEM 3. A student whirls a bucket of water in a vertical circle of radius 0.800 m. Determine the minimum a) speed and b) frequency of rotation required for the bucket to completely negotiate the top of the loop without water spilling out of the bucket.

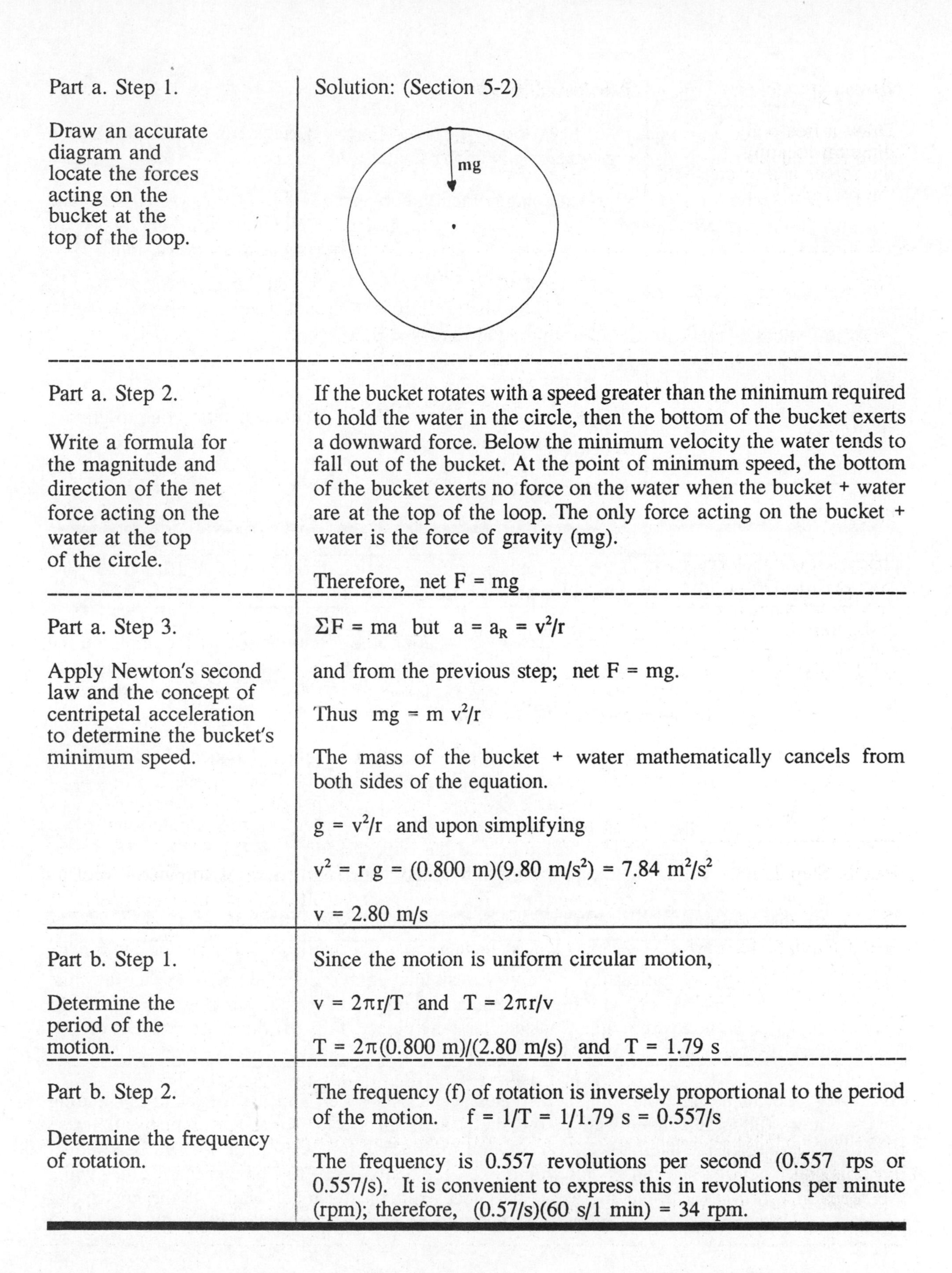

Part a. Step 1. Draw an accurate diagram and locate the forces acting on the bucket at the top of the loop.	Solution: (Section 5-2) mg
Part a. Step 2. Write a formula for the magnitude and direction of the net force acting on the water at the top of the circle.	If the bucket rotates with a speed greater than the minimum required to hold the water in the circle, then the bottom of the bucket exerts a downward force. Below the minimum velocity the water tends to fall out of the bucket. At the point of minimum speed, the bottom of the bucket exerts no force on the water when the bucket + water are at the top of the loop. The only force acting on the bucket + water is the force of gravity (mg). Therefore, net $F = mg$
Part a. Step 3. Apply Newton's second law and the concept of centripetal acceleration to determine the bucket's minimum speed.	$\Sigma F = ma$ but $a = a_R = v^2/r$ and from the previous step; net $F = mg$. Thus $mg = m\, v^2/r$ The mass of the bucket + water mathematically cancels from both sides of the equation. $g = v^2/r$ and upon simplifying $v^2 = r\, g = (0.800 \text{ m})(9.80 \text{ m/s}^2) = 7.84 \text{ m}^2/\text{s}^2$ $v = 2.80$ m/s
Part b. Step 1. Determine the period of the motion.	Since the motion is uniform circular motion, $v = 2\pi r/T$ and $T = 2\pi r/v$ $T = 2\pi(0.800 \text{ m})/(2.80 \text{ m/s})$ and $T = 1.79$ s
Part b. Step 2. Determine the frequency of rotation.	The frequency (f) of rotation is inversely proportional to the period of the motion. $f = 1/T = 1/1.79 \text{ s} = 0.557/\text{s}$ The frequency is 0.557 revolutions per second (0.557 rps or 0.557/s). It is convenient to express this in revolutions per minute (rpm); therefore, $(0.57/\text{s})(60 \text{ s}/1 \text{ min}) = 34$ rpm.

Newton's Universal Law of Gravitation

The weight of an object placed on or very near the Earth's surface can be determined by using the equation weight = mg where g is approximately 9.8 m/s^2.

In general, the force (F) of gravitational attraction between any two objects of mass m_1 and mass m_2, their centers separated by a distance r, is given by **Newton's universal law of gravitation**: $F = G\, m_1\, m_2/r^2$ where $G = 6.67 \times 10^{-11}$ N m^2/kg^2 (the universal constant).

Gravity Near the Earth's Surface

For an object of mass m on or very near the Earth's surface,

$mg = G\, m\, m_e/r_e^2$ and $g = G\, m_e/r_e^2$

where m_e is the mass of the Earth and r_e the radius of the Earth. The acceleration due to gravity at some distance from the Earth can be determined by calculating the "effective" value of g. In order to determine the effective value of g, the radius of the Earth is replaced by the value for the radial distance between the object and the center of the Earth.

TEXTBOOK QUESTION 11. Does an apple exert a gravitational force on the Earth? If so, how large a force? Consider an apple a) attached to a tree and b) falling.

ANSWER: According to the universal law of gravitation, the force exerted by the apple on the Earth must be equal in magnitude to the force exerted by the Earth on the apple. According to Newton's third law, the forces must be equal but in opposite directions. The Earth pulls the apple downward while the apple pulls the earth upward.

In part a, the apple is attached to the tree and the apple's weight is balanced by an upward force exerted by the tree branch. However, it still exerts an upward force on the Earth.

In part b, the apple is falling and the net force on the apple equals the gravitational force exerted by the Earth. The apple exerts an equal but opposite force on the Earth. However, based on Newton's second law, the upward acceleration of the Earth is negligible because of its enormous mass.

TEXTBOOK QUESTION 21. Astronauts who spend long periods in outer space could be adversely affected by weightlessness. One way to simulate gravity is to shape the spaceship like a cylindrical shell that rotates, with the astronauts walking on the inside surface (see fig. 5-32). Explain how this simulates gravity. Consider (a) how objects fall, (b) the force we feel on our feet, and (c) any other aspects of gravity you can think of.

ANSWER: Let us assume that the shell rotates at constant speed and the astronaut is standing on the inner wall. The inner wall of the shell exerts an inward force, i.e. centripetal force, toward the center of the shell. Based on Newton's third law, for every action, there is an equal but opposite reaction. The inner wall would exert the action force on the astronaut's feet while the reaction force is the force the astronaut's feet exert on the inner wall. If the centripetal acceleration equals the gravitational acceleration experienced on Earth, then the centripetal force would equal the astronaut's weight on Earth.

If the astronaut dropped something, e.g., this book, the book would travel tangent to the circle in which it is traveling until it strikes the inner wall. However, as on Earth, the astronauts would see the book drop vertically downward towards his feet. This is because the astronaut is traveling approximately in the same circle as the book and at the same rate.

In a weightless environment, bones lose calcium and muscles atrophy. As a result, today's astronauts must spend several hours each day in vigorous exercise. The centripetal force provided by a rotating cabin would provide the artificial gravity needed to prevent physiological changes.

Another advantage would be the way an astronaut drinks liquids. Astronauts now must keep liquids in closed, squeezeable containers. This is because the liquids tend to float out of an open cup and form small droplets that fill the cabin. The astronaut could drink liquids from an open cup or container but, as on Earth, he would still have to careful not to knock the over the container.

EXAMPLE PROBLEM 4. On December 24, 1968, the Apollo VIII command module became the first manned vehicle to go into orbit above the surface of the Moon. Assuming that the orbit was approximately circular and the vehicle was 110 km (69 miles) above the lunar surface, determine the a) orbital velocity of the spacecraft and b) period of the motion.
Note: $m_{Moon} = 7.40 \times 10^{22}$ kg, $r_{Moon} = 1.74 \times 10^6$ m, and 110 km = 1.10×10^5 m.

Part a. Step 1. Draw a diagram of the spacecraft's orbit about the Moon.	Solution: (Sections 5-7 and 5-8) 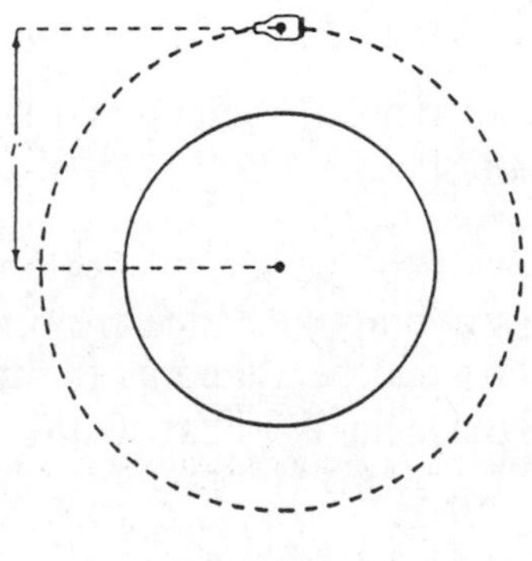
Part a. Step 2. Determine the radius of the orbit as measured from the center of the Moon.	$r = r_{Moon}$ + altitude above surface $= 1.74 \times 10^6$ m + 1.10×10^5 m $r = 1.85 \times 10^6$ m
Part a. Step 3. What force holds the spacecraft in its circular orbit about the Moon?	Lunar gravity is the only force acting on the spacecraft and this force causes it to travel in uniform circular motion. $F = G\,m\,M/r^2$

Part a. Step 4. To solve for the orbital velocity, apply Newton's second law and the concept of centripetal acceleration.	$\Sigma F = F_{grav} = m\, a_R$ $G\, m\, M/r^2 = m\, v^2/r$ The mass of the orbiting object (m) appears on both sides of the equation and cancels algebraically. $v^2 = G\, M/r$ $= [(6.67 \times 10^{-11}\ N\ m^2/kg^2)(7.4 \times 10^{22}\ kg)]/(1.85 \times 10^6\ m)$ $v^2 = 2.7 \times 10^6\ m^2/s^2$ $v = 1.6 \times 10^3$ m/s; approximately 3600 mph
Part b. Step 1. Determine the period of the motion.	$v = 2\pi r/T$ $1.6 \times 10^3\ m/s = 2\ \pi\ (1.85 \times 10^6\ m)/T$ $T = 7.12 \times 10^3$ s or 1.97 hours

Kepler's Laws and Newton's Synthesis

Johannes Kepler used data collected by Tycho Brahe to empirically derive three laws for the motion of the planets about the Sun. These laws are now referred to as Kepler's laws of planetary motion and can be summarized as follows:

First law: The path of each planet about the Sun is an ellipse with the Sun at one of the focal points of the ellipse.

Second law: Each planet moves so that an imaginary line drawn from the Sun to the planet sweeps out equal areas in equal periods of time.

Third law: The ratio of the square of the period (T^2) of any planet revolving about the Sun to the ratio of the cube of the planet's average distance from the Sun (r^3) is a constant, i.e., T^2/r^3 = constant. Also, the third law can be stated as the ratio of the squares of the periods of any two planets revolving about the Sun is equal to the ratio of the cubes of their average distances from the Sun, i.e., $T_1^2/T_2^2 = r_1^3/r_2^3$.

Newton was able to mathematically derive Kepler's laws by using his laws of motion and universal law of gravitation. Thus Kepler's laws, which were derived from empirical data, could be used to support Newton's universal law of gravitation.

TEXTBOOK QUESTION 23. The Earth moves faster in its orbit around the Sun in January than in July. Is the Earth closer to the Sun in January or in July? Explain. [Note: This not much of a factor in producing the seasons - the main factor is the tilt of the Earth's axis relative to the plane of its orbit.]

ANSWER: Based on Kepler's second law, the Earth moves faster in its orbit in January when it is closer to the Sun. The effect on the seasons due to the variation in the distance is minimal compared to the effect of the tilt of the Earth on its axis.

In July, the northern hemisphere is tilted toward the Sun. The number of hours of daylight is greater and the northern hemisphere receives more direct rays of the Sun. In January, the northern hemisphere is tilted away from the Sun. The Sun's rays are less direct and there are fewer hours of daylight. The combination of number of hours of daylight and the angle of the Sun's rays is the major cause of the variation in the seasons.

EXAMPLE PROBLEM 5. The mean distance from the Earth to the Sun is 1.496×10^8 km and the period of its motion about the Sun is 1.00 year. The period of Jupiter's motion about the Sun is observed to be 11.86 years. Determine the mean distance from the Sun to Jupiter.

Part a. Step 1.	Solution: (Section 5-8)
Complete a data table.	$T_1 = 11.86$ y $r_1 = ?$
	$T_2 = 1.00$ y $r_2 = 1.496 \times 10^8$ km
Part a. Step 2.	$T_1^2/T_2^2 = r_1^3/r_2^3$
Use Kepler's third law to solve for the distance.	$(11.86\ \text{y})^2/(1.00\ \text{y})^2 = r_1^3/(1.496 \times 10^8\ \text{km})^3$
	$r_1^3 = (11.86)^2(1.496 \times 10^8\ \text{km})^3 = 4.70 \times 10^{26}\ \text{km}^3$
	$r_1 = 7.8 \times 10^8$ km or approximately 480 million miles.

Types of Forces in Nature

Physicists now recognize only four different forces in nature. These are the gravitational force, electromagnetic force, strong nuclear force, and weak nuclear force. The electromagnetic force refers to electrical and magnetic forces. The strong nuclear force and weak nuclear force operate at the level of the atomic nucleus.

The ordinary pushes, pulls, and frictional forces that have been considered in previous chapters are referred to as "contact forces" because they involve objects that come into contact with one another, e.g., a boy pulling a wagon. According to modern understanding, contact forces are due to the electromagnetic force.

PROBLEM SOLVING SKILLS

Problems involving centripetal acceleration relate to Newton's second law of motion. Therefore, the steps followed in Chapter 4 of the textbook can be applied to this chapter. The following is an outline of the steps to be followed in attempting to solve problems related to uniform circular motion and centripetal acceleration:

1. Draw an accurate diagram locating each of the forces acting on the object in uniform circular motion. Take note of the force(s) that is/are causing the object to travel in a curved path.
2. Determine the magnitude and direction of the net force acting on the object.

3. Complete a data table based on information both given and implied in the statement of the problem.
4. Use Newton's second law and the concept of centripetal acceleration to solve the problem.

For problems involving Newton's universal law of gravitation:

1. Complete a data table based on information both given and implied in the problem.
2. Use the universal law of gravitation and, if necessary, Newton's second law and the concept of centripetal acceleration to solve the problem.

For problems involving Kepler's third law:

1. Complete a data table based on information both given and implied in the problem.
2. Use Kepler's third law to solve the problem.

SOLUTIONS TO SELECTED TEXTBOOK PROBLEMS

TEXTBOOK PROBLEM 7. A ball on the end of a string is revolved at a uniform rate in a vertical circle of radius 72.0 cm, as shown in Fig. 5-33. If its speed is 4.00 m/s and its mass is 0.300 kg, calculate the tension in the string when the ball is (a) at the top of its path and (b) at the bottom of its path.

Part a. Step 1. Fig. 5-33 locates the forces acting on the ball at the top and bottom of the loop.	Solution: (Section 5-2) 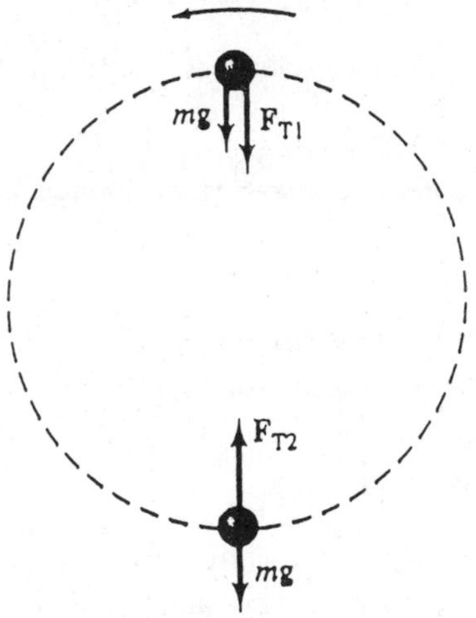 mg F_{T1} F_{T2} mg
Part a. Step 2. Complete a data table using information both given and implied.	r = 72.0 cm = 0.720 m v = 4.00 m/s m = 0.300 kg $\vec{F}_{T1} = ?$ $\vec{F}_{T2} = ?$
Part a. Step 3. Determine the ball's centripetal acceleration.	The centripetal acceleration can be written in terms of the radius of the circle and the ball's velocity as follows: $a_R = v^2/r = (4.00\ m/s)^2/(0.720\ m) = 22.2\ m/s^2$

Part a. Step 4. Apply Newton's second law to determine the tension in the string at the top of the loop.	$\Sigma\vec{\mathbf{F}} = m\,\vec{\mathbf{a}}$ where $\vec{\mathbf{a}} = \vec{\mathbf{a}}_R$ $F_{T1} + mg = m\,a_R$ $F_{T1} + (0.300\text{ kg})(9.80\text{ m/s}^2) = (0.300\text{ kg})(22.2\text{ m/s}^2)$ $F_{T1} + 2.94\text{ N} = 6.67\text{ N}$ $\vec{\mathbf{F}}_{T1} = 3.73$ N, downward
Part b. Step 1. Apply Newton's second law to determine the tension in the string at the bottom of the loop.	$\Sigma\vec{\mathbf{F}} = m\,\vec{\mathbf{a}}$ where $\vec{\mathbf{a}} = \vec{\mathbf{a}}_R$ $F_{T2} - mg = m\,a_R$ $F_{T2} - (0.300\text{ kg})(9.80\text{ m/s}^2) = (0.300\text{ kg})(22.2\text{ m/s}^2)$ $F_{T2} - 2.94\text{ N} = 6.67\text{ N}$ $\vec{\mathbf{F}}_{T2} = 9.61$ N, upward

TEXTBOOK PROBLEM 10. How large must the coefficient of static friction be between the tires and the road if a car is to round a level curve at of radius 85 m at a speed of 95 km/h?

Part a. Step 1. Draw an accurate diagram locating all of the forces acting on the car.	Solution: (Section 5-2) F_N F_{fr} $F_{gravity} = mg$
Part a. Step 2 Determine the car's speed in m/s.	(95 km/h)(1000 m/km)(1 h/3600 s) = 26.4 m/s
Part a. Step 3. Complete a data table listing information both given and implied.	The road is level; therefore, the angle of the curve is 0°. m = ? μ_s = ? $\theta = 0°$ r = 85 m v = 26.4 m/s

Part a. Step 4.

Use Newton's second law, the frictional force, and the formula for centripetal acceleration to determine the coefficient of friction.

The car is held in the circle by a frictional force which provides the centripetal acceleration. At the point where the car tends to slide off the circle, the force required to hold the box in position "just" equals the maximum possible value of the frictional force.

$\Sigma F = m\,a$ and $a = v^2/r$

$F_{fr} = \mu_s F_N$ where $F_N = mg \cos \theta = mg \cos 0° = mg$

$F_{fr} = \mu_s\, mg$

The frictional force holds the car in the circle; therefore,

$\Sigma F = F_{fr}$

$m\,v^2/r = \mu_s\, mg$

Note: the mass cancels from both sides of the equation.

$\mu_s\, g = v^2/r$

$\mu_s = v^2/(rg) = (26.4 \text{ m/s})^2/[(85.0 \text{ m})(9.8 \text{ m/s}^2)]$

$\mu_s = 0.84$

TEXTBOOK PROBLEM 15. How many revolutions per minute would a 15 m diameter Ferris wheel need to make for the passengers to feel "weightless" at the topmost point?

Part a. Step 1.

Use Fig. 5-9 in the textbook to determine the direction of the net force acting on the rider at the top of the loop.

Solution: (Section 5-2)

If the Ferris wheel rotates at a speed below the minimum speed to feel weightless, then there are two forces acting on the rider. The seat exerts an upward force (F_N) on the rider while gravity provides the downward force.

At the minimum speed to feel "weightless", gravity is the only force acting on the rider. The centripetal force is due to the rider's weight.

$\Sigma \vec{\mathbf{F}} = m\,\vec{\mathbf{g}}$

Part a. Step 2.

Determine the minimum speed in order to produce "weightlessness."

Since the Ferris wheel is traveling in a circle, the net force equals the centripetal force; therefore,

$m\,v^2/r = mg$ (Note that the mass cancels)

$v^2 = r\,g$ where $r = (15.0 \text{ m})/2 = 7.5 \text{ m}$

$v^2 = (7.5 \text{ m}))(9.8 \text{ m/s}^2) = 74 \text{ m}^2/\text{s}^2$

$v = 8.6 \text{ m/s}$

Part a. Step 4. Determine the period of motion of the Ferris wheel.	$v = 2\pi\, r/T$ where T = period of the motion. $T = 2\pi\, r/v = 2\,\pi\,(7.5\text{ m})/(8.6\text{ m/s}) = 5.5\text{ s}$
Part a. Step 5. Determine the frequency of rotation in revolutions per minute (rpm).	$f = 1/T$ $f = [1/(5.50\text{ s})] = 0.18$ revolutions per second (0.18 revolutions/sec)[(60 s/1 min)] = 11 rpm

TEXTBOOK PROBLEM 18. In a "Rotor-ride" at a carnival, people are rotated in a cylindrically walled "room" (Fig. 5-35). The room radius is 4.6 m, and the rotation frequency is 0.50 revolution per second when the floor drops out. What is the minimum coefficient of static friction so that the people will not slip down? People on this ride say they were "pressed against the wall." Is there really an outward force pressing them against the wall? If so, what is its source? If not what is the proper description of their situation (besides scary)? [Hint: First draw the free-body diagram for a person.]

Part a. Step 1. Draw a free-body diagram locating the forces acting on the rider.	Solution: (Section 5-2) F_{fr}, F_N, mg
Part a. Step 2. Determine the tangential speed of a person standing along the wall.	$v = 2\,\pi\, r/T = 2\pi\, r\, f$ $v = 2\,\pi\,(4.6\text{ m})(0.50\text{ rev/s})$ $v = 14\text{ m/s}$
Part a. Step 3. What forces must balance if the rider does not slip down the wall?	The rider does not slip down the wall; therefore, the upwardly directed frictional force must be equal but opposite to the rider's weight. $\vec{\mathbf{F}}_{fr} = m\,\vec{\mathbf{g}}$
Part a. Step 4. What force provides the radial acceleration?	The net force on the rider is due to the normal force the wall exerts on the rider. This normal force is directed toward the center of circular motion and provides the centripetal or radial acceleration experienced by the rider. $F_N = m\, v^2/r$

Part a. Step 5. Use Newton's second law and the concept of centripetal acceleration to solve for the minimum value for the coefficient of friction.	$F_{fr} = \mu_s F_N$ where $F_{fr} = mg$ and $F_N = m a_c = m v^2/r$ $mg = \mu_s m v^2/r$ Note: the rider's mass algebraically cancels. $g = \mu_s v^2/r$ and rearranging gives $\mu_s = r g/v^2 = (4.6 \text{ m})(9.8 \text{ m/s}^2)/(14 \text{ m/s})^2$ $\mu_s = 0.23$

There is no outward force "pressing" the rider against the wall. The rider tends to move tangent to the circle in which the room is traveling. The wall provides the centripetal force needed to keep the passenger moving in a circle. Based on Newton's third law, while the wall exerts an inward force on the rider, the rider exerts an outward force on the wall.

TEXTBOOK PROBLEM 43. Calculate the speed of a satellite moving in a stable circular orbit about the Earth at a height of 3600 km.

Part a. Step 1. Draw a diagram of the satellite's orbit about the Earth.	Solution: (Section 5-8) 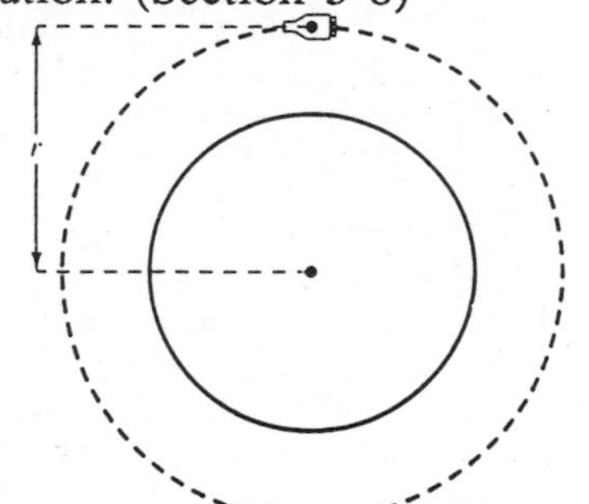
Part a. Step 2. Determine the radius of the orbit as measured from the center of the Earth. Note: 1 km = 1000 m.	The mass and radius of the Earth are given in inside front cover of the textbook. $m_{Earth} = 5.97 \times 10^{24}$ kg, $r_{Earth} = 6.38 \times 10^6$ m, $r = r_{Earth}$ + altitude above surface $r = 6.38 \times 10^6 \text{ m} + (3600 \text{ km})[(1000 \text{ m})/)1 \text{ km})]$ $r = 6.38 \times 10^6 \text{ m} + 3.60 \times 10^6 \text{ m} = 9.98 \times 10^6 \text{ m}$
Part a. Step 3. What forces are acting on the spacecraft?	The Earth's gravity is the only force acting on the spacecraft and this force causes it to travel in uniform circular motion. $F = G m M/r^2$

Part a. Step 4.

To solve for the orbital velocity, apply Newton's second law and the concept of centripetal acceleration.

$\vec{F}_{grav} = m\,\vec{a}_R$

$G\,m\,M/r^2 = m\,v^2/r$

The mass of the orbiting object (m) appears on both sides of equation and cancels algebraically.

$v^2 = G\,M/r$

$= [(6.67 \times 10^{-11}\ N\ m^2/kg^2)(5.97 \times 10^{24}\ kg)]/(9.98 \times 10^6\ m)$

$v^2 = 3.99 \times 10^7\ m^2/s^2$

$v = 6.32 \times 10^3$ m/s; approximately 14,100 mph

TEXTBOOK PROBLEM 53. What will the spring scale read for a 55 kg student in an elevator that moves (a) upward with a constant speed of 6.0 m/s, (b) downward with a constant speed of 6.0 m/s, (c) upward with an acceleration of 0.33 g, (d) downward with an acceleration of 0.33 g, and (e) in free fall?

Part a. Step 1.

Draw a diagram locating the forces acting on the student.

Solution: (Sections 5-8)

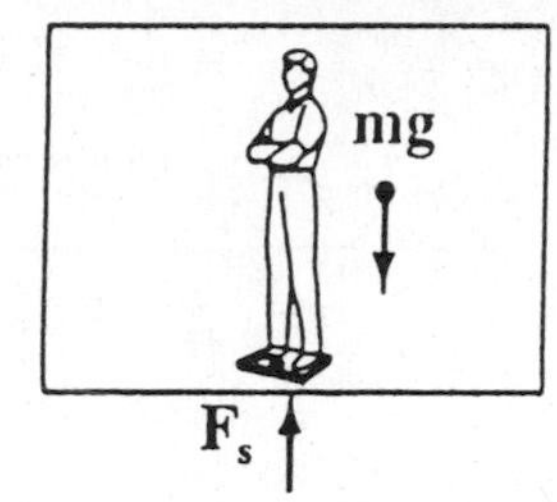

Part a. Step 2.

Determine the reading on the scale when the elevator is traveling upward at a constant speed of 6.0 m/s.

The velocity is constant; therefore, the rate of acceleration is zero and the net force must equal zero. The student's apparent weight as read by the scale equals the actual weight.

$\Sigma F = m\,a$ and $F_s - mg = m\,a$ but $a = 0\ m/s^2$

$F - (55\ kg)(9.8\ m/s^2) = (55\ kg)(0\ m/s^2)$

$F = 540\ N$

Part b. Step 1.

Determine the reading on the scale when the elevator is traveling downward at a constant speed of 6.0 m/s.

The velocity is constant; therefore, the rate of acceleration is zero and the net force must equal zero. The student's apparent weight as read by the scale equals the actual weight.

$F = m\,a$ and $F_s - mg = m\,a$ but $a = 0\ m/s^2$

$F - (55\ kg)(9.8\ m/s^2) = (55\ kg)(0\ m/s^2)$

$F = 540\ N$

Part c. Step 1. Determine the scale reading when the elevator is accelerating upward at 0.33 g.	Since the acceleration is upward, then F_s is greater than mg and the net force is F_s - mg, directed upward. The student's apparent weight as determined by the scale is greater than the actual weight. Apply Newton's second law and determine the scale reading (F_s). $F = m a$ and $F_s - m g = m a$ $F_s - (55\ kg)(9.8\ m/s^2) = (55\ kg)[(0.33)(9.8\ m/s^2)]$ $F_s - 540\ N = 180\ N$ $F_s = 720\ N$
Part d. Step 1. Determine the scale reading when the elevator is accelerating downward at 0.33 g.	Since the acceleration is downward, then F_s is less than mg and the net force is mg - F_s, directed downward. The student's apparent weight as determined by the scale is less than the actual weight. Apply Newton's second law and determine the scale reading (F_s). $F = m a$ and $mg - F_s = m a$ $(55\ kg)(9.8\ m/s^2) - F_s = (55\ kg)[(0.33)(9.8\ m/s^2)]$ $540\ N - F_s = 180\ N$ $F_s = 360\ N$
Part e. Step 1. Determine the reading on the scale when it is in free fall.	When the elevator is in free fall, its downward rate of acceleration equals 9.8 m/s^2. $\Sigma F = m a$ and $mg - F_s = m a$ $(55\ kg)(9.8\ m/s^2) - F_s = (55\ kg)(9.8\ m/s^2)$ $F_s = 0\ N$ The spring scale will not exert a force on the student. During free fall, the student will experience "apparent" weightlessness.

CHAPTER 6

WORK AND ENERGY

OBJECTIVES

After studying the material of this chapter, the student should be able to:

- distinguish between the scientific definition of work as compared to the colloquial definition.
- write the definition of work in terms of force and displacement and calculate the work done by a constant force when the force and displacement vectors are at an angle.
- use graphical analysis to calculate the work done by a force that varies in magnitude.
- define types of mechanical energy and give examples of types of energy that are not mechanical.
- state the work-energy theorem and apply the theorem to solve problems.
- distinguish between a conservative and a nonconservative force and give examples of each type of force.
- state the law of conservation of energy and apply the law to problems involving mechanical energy.
- define power in the scientific sense and solve problems involving work and power.

KEY TERMS AND PHRASES

work is done on an object when the energy of the object changes. The work done equals the product of the net force (F), the displacement (d), and the cosine of the angle between the force vector and the displacement vector.

joule (J) is the unit associated with work and energy. $1 \text{ J} = 1 \text{ N m} = 1 \text{ kg m/s}^2$.

kinetic energy (KE) is energy due to an object's motion. Translational kinetic energy of an object depends on the object's mass and the square of its speed.

work-energy theorem states that the net work done on an object is equal to its change in kinetic energy.

gravitational potential energy (PE) is energy stored by an object due to its position. Near the surface of the Earth, the gravitational potential energy relative to some reference point, e.g., the ground, is given by the product of the object's weight and the height above the reference level.

elastic potential energy is energy that results from an object being stretched or compressed (e.g., a spring), or twisted as in case of a wire. The change in a spring's elastic potential energy as it is stretched (or compressed) is related to the force (F) required to stretch (or compress) the spring and the distance that the spring is stretched (or compressed).

conservative force does the same amount of work independent of the path taken between two points. An example of a conservative force is gravity. The work done equals the change in potential energy ($W = \Delta PE = mg\, \Delta y$) and depends only on the initial and final positions above the ground and not on the path taken.

nonconservative force results in different amounts of work being expended in moving an object. The net work required depends on the path taken moving the object between points. For example, the work done in sliding a box of books against friction from one end of a room to the other depends on the path taken.

mechanical energy refers to an object's kinetic energy and potential energy.

law of conservation of energy states that energy is neither created nor destroyed. Energy can be transformed from one kind to another, but the total amount remains constant.

power is the rate at which work is done and is measured in watts where 1 watt = 1 J/s.

SUMMARY OF MATHEMATICAL FORMULAS

work	$W = F\, d \cos \theta$	The work done depends on the net force, displacement, and the angle between the force vector and the displacement vector.
kinetic energy	$KE = \frac{1}{2} m v^2$	Kinetic energy is energy of motion. Kinetic energy depends on an object's mass and the square of its speed.
work-energy theorem	$W = \frac{1}{2} m v_f^2 - \frac{1}{2} m v_i^2$	The work-energy theorem states that the net work done on an object is equal to its change in kinetic energy.
gravitational potential energy	$PE = m g y$	For an object near the surface of the Earth, the gravitational potential energy depends on its weight and its height above a reference level.
spring or elastic potential energy	$PE = \frac{1}{2} k x^2$	Elastic potential energy depends on the force constant of the spring (k) and the displacement from equilibrium position (x).
power	$P = W/t$ or $P = F \bar{v} \cos \theta$	Power is the rate at which work is done, or for an object traveling at constant speed, the power expended is related to the net force and the average speed.

CONCEPT SUMMARY

Work and Energy

The **work** (W) done by a force (F) that is constant in both magnitude and direction is given by the following equation:

$$W = F\, d \cos \theta$$

where d is the displacement of the object and θ is the angle between the force vector and the displacement vector.

Work is a scalar quantity. Since work is the product of a force and displacement, the unit of work is newton meter. The newton meter has been named the **joule** (J), and 1 J = 1 N m.

If the force acting on an object varies in magnitude and/or direction during the object's displacement, graphical analysis can be used to determine the work done. $F \cos \theta$ is plotted on the y axis and the distance through which the object moves is plotted on the x axis. The work done is represented by the area under the curve. The sum of the complete and partial blocks under the curve is determined and the work done equals the product of the work represented by one block and the total number of blocks.

A moving object is said to have **kinetic energy** (KE). The translational kinetic energy of an object depends on the object's mass and the square of its speed. The formula for kinetic energy is

$$KE = \tfrac{1}{2} m v^2$$

When a force does work on an object, the energy of the object will be changed by an amount equal to the amount of work done. The work done is positive if the energy of the object increases, for example, a car accelerating from rest. The work done is negative if the energy of the object decreases; for example, the kinetic energy of a sliding object on a rough surface is gradually lost as the object comes to a halt. In this case the kinetic energy is dissipated into the form of heat energy. The **work-energy theorem** states that the net work done on an object is equal to its change in kinetic energy, i.e., $W = \Delta KE = \tfrac{1}{2} m v_2^2 - \tfrac{1}{2} m v_1^2$.

Gravitational potential energy (PE) is energy stored by an object due to its position. The formula for the gravitational potential energy of an object relative to some reference point, e.g., the ground, is given by

$PE = m g y$ where y = the difference in height and the reference level

If an object of mass m is raised from an initial height y_1 and raised to a height y_2 above its original position, the change in gravitational potential energy is given by

$$\Delta PE = m g y_2 - m g y_1$$

Elastic potential energy is energy that results from an object being stretched or compressed (e.g., a spring), or twisted as in case of a wire. The force (F) required to stretch or compress a spring is given by $F = k x$ where k is the force constant measured in newtons/meter (N/m) and x is the stretch or compression. The change in a spring's elastic potential energy as it is stretched or compressed is given by

$\Delta PE = \frac{1}{2} k x_2^2 - \frac{1}{2} k x_1^2$ where x_1 = the initial displacement from the equilibrium
and x_2 = the final displacement from equilibrium

TEXTBOOK QUESTION 2. Can a centripetal force ever do work on an object? Explain.

ANSWER: For an object in traveling in *circular* motion, the centripetal force *never* does work. The centripetal force is directed toward the center of the circle while the displacement vector is tangent to the circle. Therefore, the centripetal force vector is always perpendicular to the displacement vector ($\theta = 90°$). However, cos 90° = 0; therefore, $W = F\, d \cos 90° = 0$.

An alternative answer can be given from the work-energy theorem. The centripetal force changes the direction of motion of an object traveling in a circle without changing the object's speed. If the object's speed does not change, then its kinetic energy does not change and no work was done on the object. Note: if the motion is elliptical instead of circular, then θ is not equal to 90° and work is done on the object.

EXAMPLE PROBLEM 1. A student exerts a horizontal force of 20.0 N with her hand and pushes a 10.0 kg box a distance of 2.00 m across a frictionless floor. Calculate the a) magnitude of the work done by the student and b) velocity of the box at the 2.00 m mark.

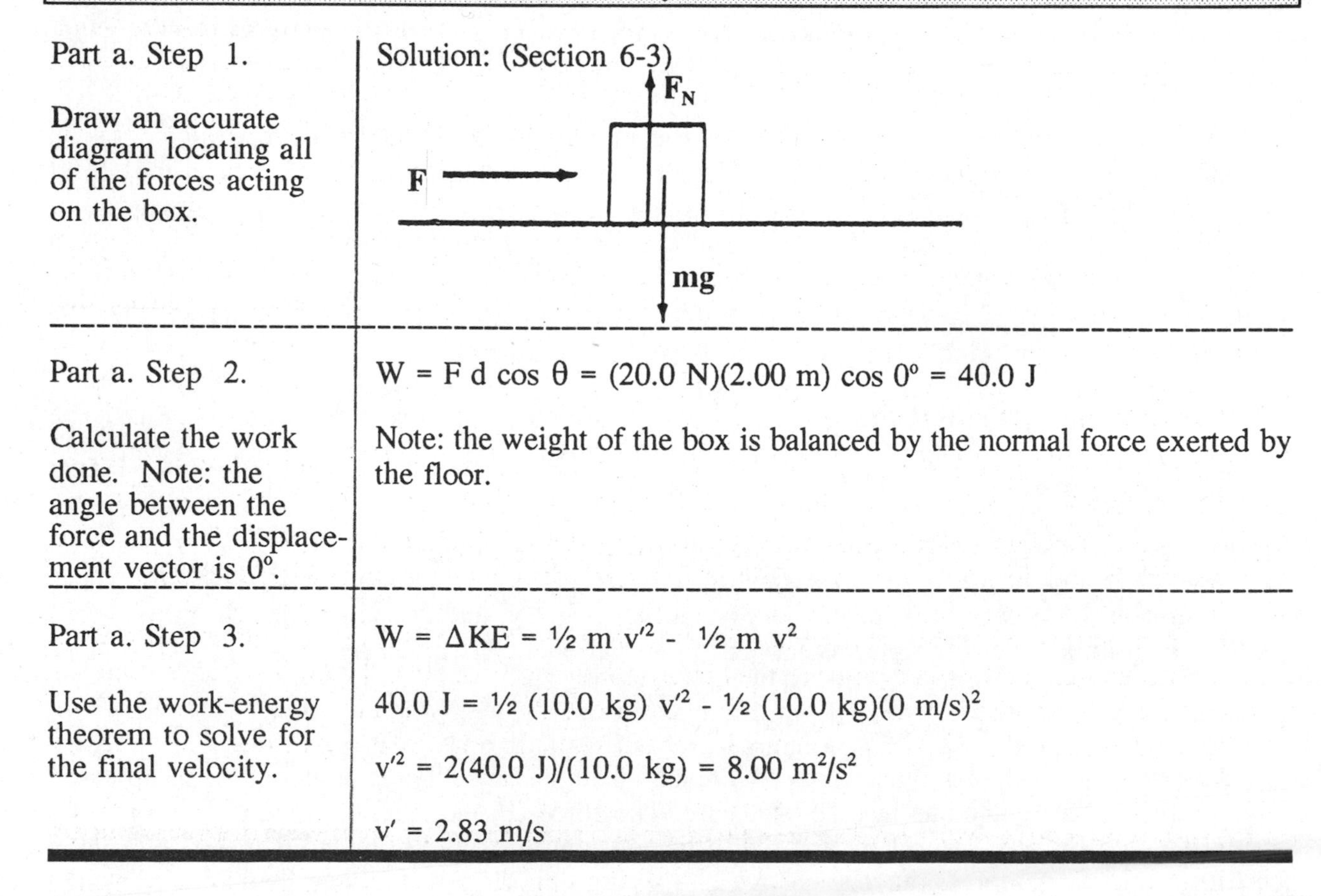

Part a. Step 1. Draw an accurate diagram locating all of the forces acting on the box.	Solution: (Section 6-3)
Part a. Step 2. Calculate the work done. Note: the angle between the force and the displacement vector is 0°.	$W = F\, d \cos\theta = (20.0\ N)(2.00\ m) \cos 0° = 40.0\ J$ Note: the weight of the box is balanced by the normal force exerted by the floor.
Part a. Step 3. Use the work-energy theorem to solve for the final velocity.	$W = \Delta KE = \frac{1}{2} m v'^2 - \frac{1}{2} m v^2$ $40.0\ J = \frac{1}{2} (10.0\ kg) v'^2 - \frac{1}{2} (10.0\ kg)(0\ m/s)^2$ $v'^2 = 2(40.0\ J)/(10.0\ kg) = 8.00\ m^2/s^2$ $v' = 2.83\ m/s$

Conservative and Nonconservative Forces

The work done by a **conservative force** depends only on the initial and final position of the object acted upon. An example of a conservative force is gravity. The work done equals the change in potential energy ($W = \Delta PE = mg\ \Delta y$) and depends only on the initial and final positions above the ground and *not* on the path taken.

Friction is a **nonconservative force** and the work done in moving an object against a nonconservative force depends on the path. For example, the work done in sliding a box of books against friction from one end of a room to the other depends on the path taken.

Law of Conservation of Energy

The **law of conservation of energy** states that energy is neither created nor destroyed. Energy can be transformed from one kind to another, but the total amount remains constant. For mechanical systems involving conservative forces, the total **mechanical energy** equals the sum of the kinetic and potential energies of the objects that make up the system.

$E = KE + PE$

An example of the law of conservation of energy involving a mechanical system is a ball thrown vertically upward. Neglecting air resistance, kinetic energy is gradually transformed into potential energy and then back into kinetic energy. At each point of the motion the sum of the kinetic energy and potential energy remains constant.

TEXTBOOK QUESTION 10. In Fig. 6-31, water balloons are tossed from the roof of a building, all with the same speed but with different launch angles. Which one has the highest speed on impact? Ignore air resistance.

ANSWER: When the water balloon is initially released it has both potential and kinetic energy. Assuming that air resistance is negligible and that the surface of the ground below is the zero of potential energy, the balloon's energy is all kinetic energy when it is about to strike the ground.

$PE_{building} + KE_{building} = KE_{ground}$

Each balloon's kinetic energy as it strikes the ground depends only on the total initial energy and not the angle at which it was thrown. The balloon's speed is a scalar quantity and at the ground (v_{ground}) is related to the kinetic energy KE_{ground} by the equation $KE = \frac{1}{2} m v^2$. Therefore, the balloon's speed as it is about to strike the ground is independent of the launch angle.

If air resistance is NOT negligible, then the speed does depend on the launch angle. A balloon thrown upward at some angle to the horizontal will encounter more air resistance because it travels a greater distance through the air than a balloon thrown vertically downward. Because of this, if air resistance is a factor, a balloon thrown vertically downward will have the greatest speed.

TEXTBOOK QUESTION 13. A bowling ball is hung from the ceiling by a steel wire (Fig. 6-33). The instructor pulls the ball back and stands against the wall with the ball against his nose. To avoid injury the instructor is suppose to release the ball without pushing it. Why?

ANSWER: At release, the ball has potential energy (PE) but no kinetic energy (KE). Based on the law of conservation of energy, the total energy cannot exceed the initial energy. Therefore, as the ball swings away from the instructor, the total KE + PE equals the initial PE. Assuming frictional losses are negligible, when the ball returns it will barely touch the instructor's nose before moving away again.

However, if the ball is pushed it will have both PE + KE and when it returns it will tend to reach a higher point above the ground than the point where the instructor has placed his nose. The lecturer will be hit by a moving bowling ball.

Note: as a young teacher, the author of this study guide performed this particular experiment using a student volunteer. A 2 kg cylinder attached to the ceiling by a thick string was displaced and released. Unfortunately, the string broke as the object was returning to a student's nose and the cylinder struck the student in the stomach. Since that experience, the demonstration has been performed as a "thought" experiment.

TEXTBOOK QUESTION 14. What happens to the gravitational potential energy when water at the top of a waterfall falls to the pool below?

ANSWER: If air resistance is negligible, then the gravitational potential energy and kinetic energy of the water at the top of the waterfall is converted into kinetic energy at the pool below. When the falling water strikes the pool below then it is converted into thermal energy. Therefore, the water in the pool below should be at a higher temperature than the water at the top of the waterfall.

James Joule, a British physicist, measured the temperature of the water at both the top and bottom of a high waterfall and found that the temperature at both locations was about the same. He theorized that evaporative cooling of the falling water as it mixed with the air led to essentially no change in temperature.

EXAMPLE PROBLEM 2. A stone is thrown vertically upward with an initial speed of 24.5 m/s from the edge of the roof of a building 29.4 m high. The stone just misses the edge of the building on the way down and strikes the street below. Use the law of conservation of energy to determine the a) maximum height above the ground reached by the stone and b) stone's velocity just before it strikes the ground.

Part a. Step 1.	Solution: (Sections 6-6 and 6-7)
Complete a data table.	Let point 1 represent the edge of the roof, point 2 represent maximum height, and point 3 just above ground level.
	$h_1 = 29.4$ m $\quad v_1 = 24.5$ m/s $\qquad g = 9.80$ m/s^2

	h_2 = ? v_2 = 0 m/s h_3 = 0 m v_3 = ?
Part a. Step 2. Use the law of conservation of energy to determine the maximum height reached by the stone.	$PE_1 + KE_1 = PE_2 + KE_2$ $mgh_1 + \frac{1}{2} m v_1^2 = mgh_2 + \frac{1}{2} m v_2^2$ Note: the mass cancels. $(9.80 \text{ m/s}^2)(29.4 \text{ m}) + \frac{1}{2}(24.5 \text{ m/s})^2 = (9.80 \text{ m/s}^2) h_2 + \frac{1}{2}(0 \text{ m/s})^2$ $288 \text{ m}^2/\text{s}^2 + 300 \text{ m}^2/\text{s}^2 = (9.80 \text{ m/s}^2) h_2$ $h_2 = 60.0 \text{ m}$
Part b. Step 1 Use the law of conservation of energy to determine the stone's speed just before it struck the ground.	$PE_2 + KE_2 = PE_3 + KE_3$ $mgh_2 + \frac{1}{2} m v_2^2 = mgh_3 + \frac{1}{2} m v_3^2$ Note: the mass cancels. $(9.80 \text{ m/s}^2)(60.0 \text{ m}) + \frac{1}{2}(0 \text{ m/s})^2 = (9.80 \text{ m/s}^2)(0 \text{ m}) + \frac{1}{2} v_3^2$ $588 \text{ m}^2/\text{s}^2 = \frac{1}{2} v_3^2$ $v_3 = 34.2 \text{ m/s}$

EXAMPLE PROBLEM 3. A 5.00 kg box is pushed from the top of an incline and travels down the incline with an initial speed of 10.0 m/s. The incline is 5.00 m long and the angle of the incline is 30.0°. The coefficient of friction between the box and the incline is 0.200. Determine the a) loss of potential energy and b) work done by friction. c) Use the work-energy principle to determine the velocity of the box at the bottom of the incline.

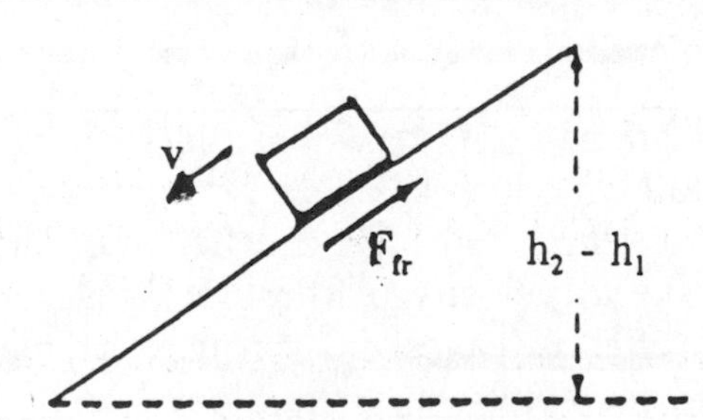

Part a. Step 1. Complete a data table.	Solution: (Sections 6-1, 6-4, and 6-9) $h_1 = (5.00 \text{ m})(\sin 30.0°) = 2.50 \text{ m}$ $h_2 = 0 \text{ m}$ $v_1 = 10.0 \text{ m/s}$ $v_2 = ?$ $F_{fr} = \mu \, mg \cos\theta = (0.200)(5.00 \text{ kg})(9.80 \text{ m/s}^2)(\cos 30.0°) = 8.49 \text{ N}$
Part a. Step 2. Determine the loss of potential energy.	$\Delta PE = mgh_2 - mgh_1 = mg(h_2 - h_1)$ $= (5.00 \text{ kg})(9.80 \text{ m/s}^2)(0 \text{ m} - 2.50 \text{ m})$ $\Delta PE = -123 \text{ J}$

Part b. Step 1.

Determine the work done by friction.

The frictional force and the direction of motion are in opposite directions; therefore, $\theta = 180°$.

$W = F\,d\cos\theta = (8.49\ N)(5.00\ m)(\cos 180°) = -42.5\ J$

The mechanical energy lost by the system equals the work done by friction.

Part c. Step 1.

Determine the total mechanical energy at the top of the incline.

The total energy at the top of the slide is the sum of the PE and KE at that point.

total energy $= (5.00\ kg)(9.80\ m/s^2)(2.50\ m) + \frac{1}{2}(5.00\ kg)(10.0\ m/s)^2$

total energy $= 123\ J + 250\ J = 373\ J$

Part c. Step 2.

Use the law of conservation of energy to solve for the speed of the box at the bottom of the incline.

The total kinetic + potential energy at the top of the final section is greater than that at the bottom by an amount equal to the mechanical energy lost due to friction.

$PE_1 + KE_1 +$ energy lost(friction) $= PE_2 + KE_2$

$373\ J + -42.5\ J = (5.00\ kg)(9.80\ m/s^2)(0\ m) + \frac{1}{2}(5.00\ kg)\,v_2^2$

$331\ J = (2.50\ kg)\,v_2^2$

$v_2^2 = 132\ m^2/s^2$

$v_2 = 11.5\ m/s$

EXAMPLE PROBLEM 3. As shown in the diagram, a 1.00 kg wooden block is on a level board and held against a spring of force constant 30.0 N/m which has been compressed 0.200 m. The block is released and propelled horizontally across the board. The coefficient of friction between the block and the board is 0.20. Determine the a) velocity of the block just as it leaves the spring and b) distance that the block travels after it leaves the spring. Assume that friction between the block and board is negligible until the point where the block leaves the spring.

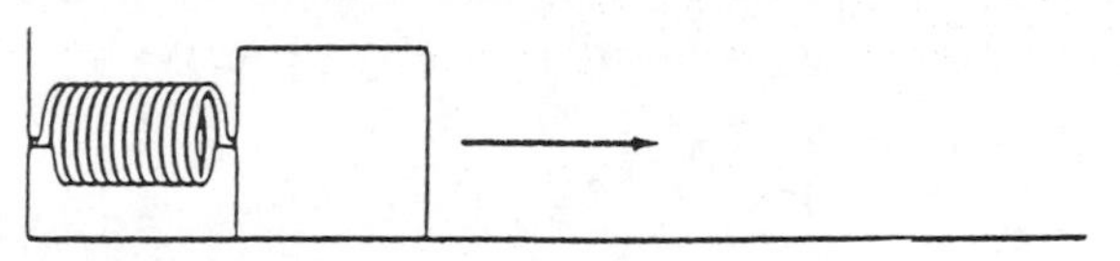

Part a. Step 1.

Use the law of conservation of energy to determine the block's speed as it leaves the spring.

Solution: (Sections 6-7 and 6-8)

$\Delta PE_{spring} = \Delta KE_{block}$

$\frac{1}{2} k x^2 = \frac{1}{2} m v^2$ and rearranging gives

$v = (k/m)^{1/2}\,x = [(30.0\ N/m)/(1.00\ kg)]^{1/2}\,(0.200\ m)$

$v = 1.10\ m/s$

Part b. Step 1.

Use the work-energy theorem to determine the distance the block slides.

work = change in kinetic energy

$F\,d\cos\theta = \frac{1}{2}\,m\,v'^2 - \frac{1}{2}\,m\,v^2$

The force acting on the block is due to friction.

$F = F_{fr} = \mu_k F_N$, where $F_N = mg$.

The frictional force is directed opposite from the direction of motion; therefore, $\cos\theta = \cos 180^\circ = -1$.

The block comes to rest; therefore, the final velocity of the block is zero ($v' = 0$).

$\mu_k\,mg\,d\cos 180^\circ = \frac{1}{2}\,m\,v'^2 - \frac{1}{2}\,m\,v^2$ The mass cancels; therefore,

$(0.200)(9.80\ \text{m/s}^2)\,d\,(-1) = 0 - \frac{1}{2}\,(1.10\ \text{m/s})^2$

$d = 0.309\ \text{m}$

Power

Power is defined as the rate at which work is done. The average power can be determined by applying the following formula:

power = work/time = energy transformed/time.

$P = W/t$

The unit of power is the **watt,** where 1 watt = 1 joule/second. The watt is related to the English unit of horsepower: 746 watts = 1 horsepower. The power output of a constant force (F) applied to an object is given by

$P = F\,\bar{v}$ where $\bar{v} = d/t$ is the average speed of the object during the time interval being considered.

EXAMPLE PROBLEM 4. A man pushes a 50.0 kg box across a level floor at a constant speed of 1.00 m/s for 10.0 s. If the coefficient of friction between the box and the floor is 0.200, determine the average power output by the man.

Part a. Step 1.

Draw an accurate diagram locating all of the forces acting on the box.

Solution: (Section 6-10)

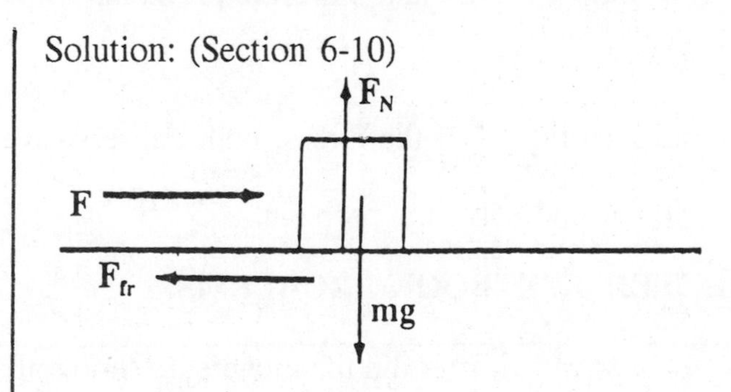

Part a. Step 2. Complete a data table.	$m = 50.0$ kg $\mu_k = 0.200$ $\bar{v} = 1.00$ m/s $t = 10$ s
Part a. Step 3. Apply Newton's second law of motion and determine the force (F) exerted by the man.	$\Sigma F = m\,a$ but $a = 0$ m/s^2 (speed is constant) $F - F_{fr} = m\,(0 \text{ m/s}^2)$ but $F_{fr} = \mu_k F_N$ $F - \mu_k F_N = 0$ and $F_N = mg$ $F - \mu_k m g = 0$ $F - (0.20)(50.0 \text{ kg})(9.80 \text{ m/s}^2) = 0$ $F = 98.0$ N
Part a. Step 4. Determine the average power output of the man.	$P = W/t = F\,d/t \cos 0°$ The speed is constant; therefore, $d/t = v$ $P = F\,v \cos 0° = (98.0 \text{ N})(1.0 \text{ m/s})$ $P = 98.0$ W

PROBLEM SOLVING SKILLS

For problems involving the work-energy theorem:

1. Draw an accurate diagram locating all of the forces, both conservative and nonconservative, acting on the object.
2. Determine the magnitude and direction of the net force acting on the object and then determine the net work done on the object.
3. Apply the work-energy theorem and solve the problem.

For problems involving the law of conservation of energy:

1. Draw an accurate diagram locating all of the forces, both conservative and nonconservative, acting on the object.
2. Apply the law of conservation of energy and solve the problem.

For problems involving power:

1. Draw an accurate diagram locating all of the forces, both conservative and nonconservative, acting on the object.
2. Apply the formulas for power and solve the problem.

SOLUTIONS TO SELECTED TEXTBOOK PROBLEMS

TEXTBOOK PROBLEM 4. How much work did the movers do (horizontally) pushing a 160 kg crate 10.3 m across a rough floor without acceleration, if the effective coefficient of friction was 0.50?

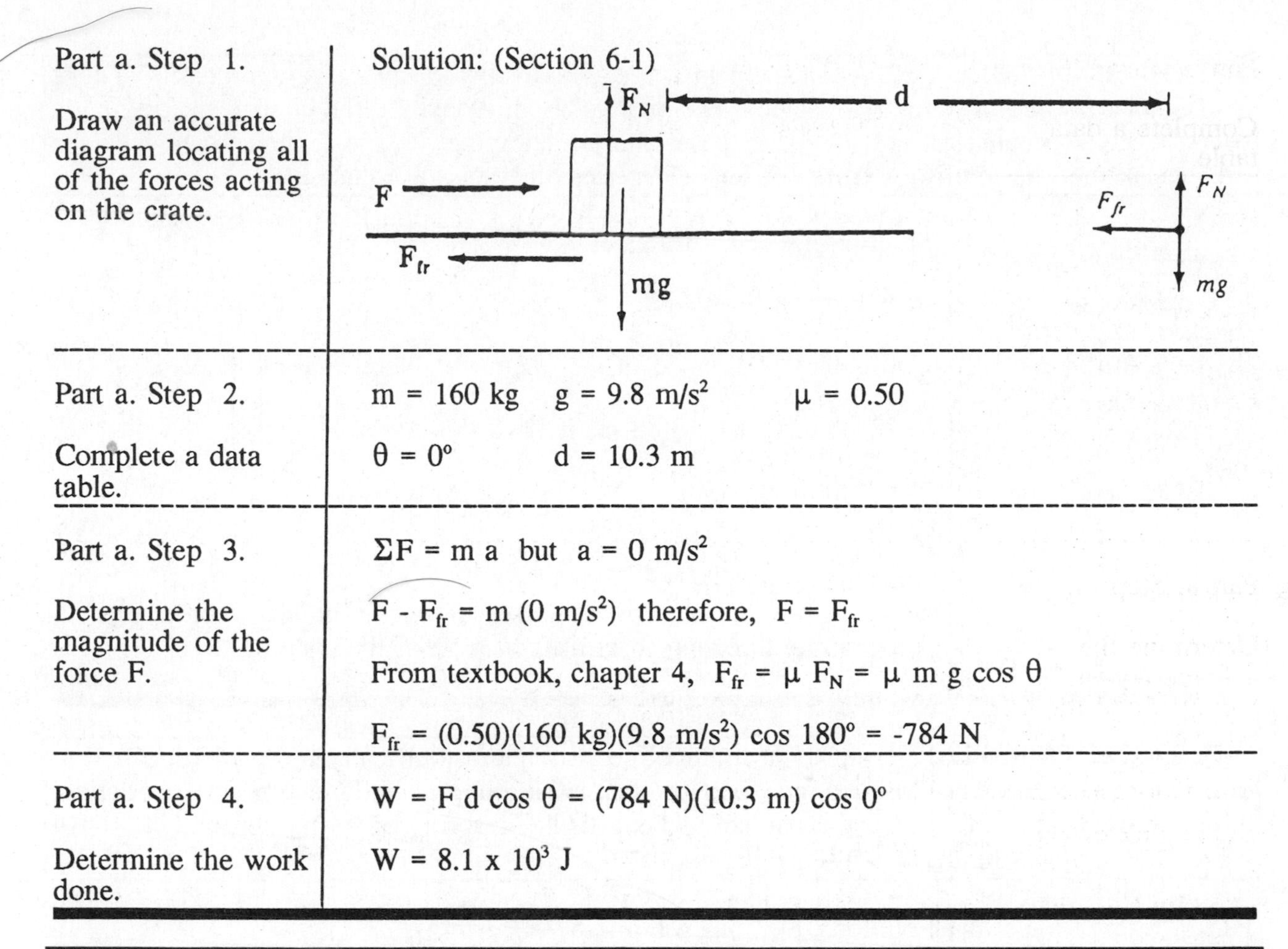

Part a. Step 1. Draw an accurate diagram locating all of the forces acting on the crate.	Solution: (Section 6-1)
Part a. Step 2. Complete a data table.	$m = 160$ kg $g = 9.8$ m/s^2 $\mu = 0.50$ $\theta = 0°$ $d = 10.3$ m
Part a. Step 3. Determine the magnitude of the force F.	$\Sigma F = m\ a$ but $a = 0$ m/s^2 $F - F_{fr} = m$ (0 m/s^2) therefore, $F = F_{fr}$ From textbook, chapter 4, $F_{fr} = \mu\ F_N = \mu\ m\ g\ \cos\theta$ $F_{fr} = (0.50)(160\text{ kg})(9.8\text{ m/s}^2)\cos 180° = -784$ N
Part a. Step 4. Determine the work done.	$W = F\ d\ \cos\theta = (784\text{ N})(10.3\text{ m})\cos 0°$ $W = 8.1 \times 10^3$ J

TEXTBOOK PROBLEM 20. A baseball (m = 140 grams) traveling 32 m/s moves a fielder's glove backward 25 cm when the ball is caught. What is the average force exerted on the ball by the glove?

Part a. Step 1. Complete a data table.	Solution: (Section 6-3) $m = 140$ grams $= 0.140$ kg $v_o = 32$ m/s $v = 0$ m/s $F = ?$ $d = 25$ cm $= 0.25$ m $\theta = 180°$
Part a. Step 2. Apply the work-energy theorem and solve for the distance.	$W = \Delta KE$ $F\ d\ \cos\theta = ½\ m\ v'^2 - ½\ m\ v^2$ $F\ (0.25\text{ m})(\cos 180°) = ½(0.140\text{ kg})(0\text{ m/s})^2 - ½(0.140\text{ kg})(32.0\text{ m/s})^2$ $-F\ (0.25\text{ m}) = 0\text{ J} - 71.7\text{ J}$ $F = 2.9 \times 10^2$ N Note: the value of θ is 180° because the force the glove exerts is directed opposite from the direction of the ball's motion.

TEXTBOOK PROBLEM 22. At an accident scene on a level road, investigators measure a car's skid mark to be 88 m long. The accident occurred on a rainy day and the coefficient of kinetic friction was estimated to be 0.42. Use these data to determine the speed of the car when the driver slammed on (and locked) the brakes. (Why does the mass not matter?)

Part a. Step 1.	Solution: (Section 6-3)
Use the work-energy theorem to determine the car's initial speed.	$W = F\ d \cos\theta = \Delta KE$ where $F = -F_{fr} = \mu_s\ m\ g \cos 180^\circ$ $\mu_s\ m\ g \cos 180^\circ\ d = ½\ m\ v'^2 - ½\ m\ v^2$ Note: mass m cancels $(0.42)(9.8\ m/s^2)(-1.0)(88\ m) = ½\ (0\ m/s)^2 - ½\ v^2$ $v^2 = 720\ m^2/s^2$ $v = 27\ m/s \approx 60$ miles per hour Note: the value of θ is 180° because the braking force acts in the opposite direction from the direction of the car's motion.

TEXTBOOK PROBLEM 37. A 65 kg trampoline artist jumps vertically upward from the top of a platform with a speed of 5.0 m/s. (a) How fast is he going as he lands on the trampoline, 3.0 m below (Fig. 6-38)? (b) If the trampoline behaves like a spring of spring constant 6.2×10^4 N/m, how far does he depress it?

Part a. Step 1.	Solution: (Sections 6-6 and 6-7)
Complete a data table. Measure all distances from the point where the artist first touches the trampoline.	Let point 1 represent the top of the platform and point 2 represent the top of the trampoline. $g = 9.8\ m/s^2$ $h_1 = 3.0\ m$ $v_1 = 5.0\ m/s$ $h_2 = 0$ $v_2 = ?\ m/s$
Part a. Step 2. Use the law of conservation of energy to determine the artist's speed as he strikes the trampoline.	$PE_1 + KE_1 = PE_2 + KE_2$ $mgh_1 + ½\ m\ v_1^2 = mgh_2 + ½\ m\ v_2^2$ Note: the mass cancels $(9.8\ m/s^2)(3.0\ m) + ½\ (5.0\ m/s)^2 = (9.8\ m/s^2)(0) + ½\ v_2^2$ $29.4\ m^2/s^2 + 12.5\ m^2/s^2 = ½\ v_2^2$ $v_2 = 9.2\ m/s$

Part b. Step 1. Complete a data table using data for points 2 and 3.	Let point 3 represent the maximum depression of the trampoline. This point is below the top of the trampoline. Therefore, let the maximum depression be represented by Y and h_3 = -Y. $g = 9.8\ m/s^2$ $v_2 = 9.2\ m/s$ $h_2 = 0\ m$ $h_3 = -Y$ $v_3 = 0\ m/s$
Part b. Step 2. Use the law of conservation of energy to determine the maximum depression of the trampoline.	$PE_2 + KE_2 = PE_{3(gravity)} + KE_3 + PE_{3(trampoline)}$ where $h_3 = -Y$, and $v_3 = 0\ m/s$, then $KE_3 = 0\ J$ $mgh_2 + ½\ m\ v_2^2 = mgh_3 + ½\ m\ v_3^2 + ½\ k\ y_3^2$ $(65.0\ kg)(9.80\ m/s^2)(0\ m) + ½\ (65.0\ kg)\ (9.2\ m/s)^2 = (65.0\ kg)(9.80\ m/s^2)(-Y) + 0\ J + ½\ (6.2 \times 10^4\ N/m)\ Y^2$ $0\ J + 2750\ J = (637\ N)(-Y) + (3.1 \times 10^4)\ Y^2$ rearranging gives $0 = (3.1 \times 10^4)\ Y^2 - 637\ Y - 2750$ Using the quadratic formula to determine the value of Y gives Y = 0.31 m or Y = -0.29 m. The positive value of Y represents the maximum depression; therefore, the correct answer is Y = 0.31 m.

TEXTBOOK PROBLEM 65. A shot-putter accelerates a 7.3 kg shot from rest to 14 m/s. If this motion takes 1.5 s, what average power was developed?

Part a. Step 1. Determine the change in the object's kinetic energy.	Solution: (Section 6-10) $\Delta KE = KE_f - KE_i$ $= ½\ m\ v_f^2 - ½\ m\ v_i^2$ $= ½\ (7.3\ kg)(14\ m/s)^2 - ½\ (7.3\ kg)(0\ m/s)^2$ $\Delta KE = 720\ J$
Part a. Step 2. Determine the average power expended by the shot-putter.	$P = W/t = \Delta KE/t$ $= (720\ J)/(1.5\ s)$ $P = 4.8 \times 10^2\ W$

TEXTBOOK PROBLEM 77. A ball is attached to a horizontal cord of length L whose other end is fixed (Fig 6-43). (a) If the ball is released, what will be its speed at the lowest point in its path? (b) A peg is located a distance h directly below the point of attachment of the cord. If h = 0.80 L, what will be the speed of the ball when it reaches the top of its circular path about the peg?

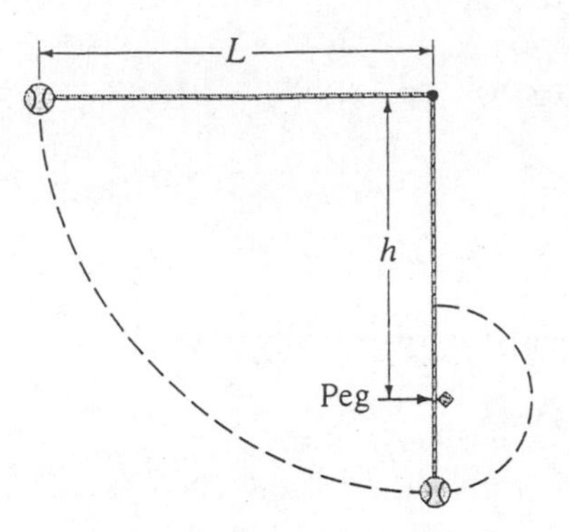

Part a. Step 1.	Solution: (Section 6-7)
Use the law of conservation of energy to solve for the ball's speed at the lowest point in its path. Hint: the ball falls a distance equal to L to reach the lowest point in its motion.	The total kinetic + potential energy at point 1 equals the total kinetic + potential energy at point 2. Note: point 1 is the release point and point 2 is the lowest point in the ball's path. $PE_1 + KE_1 = PE_2 + KE_2$ $m\ g\ h_1 + ½\ m\ v_1^2 = m\ g\ h_2 + ½\ m\ v_2^2$ Note: the mass cancels where $h_1 = L$, $v_1 = 0$ m/s, let $h_2 = 0$ m, $v_2 = ?$ $g\ L + ½\ (0\ m/s)^2 = g\ (0\ m) + ½\ v_2^2$ $v_2^2 = 2\ g\ L$ and $v_2 = (2\ g\ L)^{½}$
Part b. Step 1.	The total kinetic + potential energy at point 1 equals the total kinetic + potential energy at point 3. Note: point 3 is the top of the circular path above the peg. Note point 3 equals the diameter of the circle that the ball follows. Since the peg is 0.8L below the support point, the radius of the circle is L - 0.80 L = 0.20 L. The diameter of the circle is 2(0.20 L) = 0.40 L.
Use the law of conservation of energy speed at the top of the circular path above the peg.	$PE_1 + KE_1 = PE_3 + KE_3$ $m\ g\ h_1 + ½\ m\ v_1^2 = m\ g\ h_3 + ½\ m\ v_3^2$ Note: the mass cancels where $h_1 = L$, $v_1 = 0$ m/s, $h_3 = 0.40$ L, $v_3 = ?$ $g\ L + ½\ (0\ m/s)^2 = g\ (0.40\ L) + ½\ v_3^2$ $½\ v_3^2 = g\ L - 0.40\ g\ L = 0.60\ g\ L$ $v_3 = (1.2\ g\ L)^{½}$

CHAPTER 7

LINEAR MOMENTUM

OBJECTIVES

After studying the material of this chapter, the student should be able to:

- define linear momentum and write the mathematical formula for linear momentum from memory.
- distinguish between the unit of force and momentum.
- write Newton's second law of motion in terms of momentum.
- define impulse and write the equation that connects impulse and momentum.
- state the law of conservation of momentum and write, in vector form, the law for a system involving two or more point masses.
- distinguish between a perfectly elastic collision and a completely inelastic collision.
- apply the laws of conservation of momentum and energy to problems involving collisions between two point masses.
- define center of mass and center of gravity and distinguish between the two concepts.

KEY TERMS AND PHRASES

linear momentum ($\vec{\mathbf{p}}$) is a measure of the product of an object's mass (m) and its velocity ($\vec{\mathbf{v}}$).

law of conservation of momentum states that in any collision between two or more objects, the vector sum of the momenta before impact equals the vector sum of the momenta after impact. The total momentum remains constant, i.e., the momentum is conserved.

impulse is defined as the product of the force ($\vec{\mathbf{F}}$) acting on an object and the time (Δt) during which the force acts. The impulse equals the change in the object's momentum.

perfectly elastic collisions occur when the kinetic energy as well as the linear momentum is conserved.

inelastic collisions occur when some (or most) of the kinetic energy of the objects involved in a collision is converted into heat during impact. Problems involving this type of collision can be solved by applying the law of conservation of momentum.

center of gravity (cg) of an object is the point where the entire weight of the object can be considered to be concentrated. At the center of gravity the entire weight of the object can be balanced by a single vertical force equal to the object's weight.

center of mass (CM) is the point where all of the mass of an object can be considered to be concentrated. For almost all objects, the center of mass is at the same point in the object as the center of gravity.

translational motion is where, at any instant, the motion of all points of a moving object has the same velocity and direction of motion. This type of motion, as well as the concept of center of mass, has been implied in solving problems in previous chapters.

SUMMARY OF MATHEMATICAL FORMULAS

momentum	$\vec{\mathbf{p}} = m\,\vec{\mathbf{v}}$	Linear momentum ($\vec{\mathbf{p}}$) is defined as the product of an object's mass and its velocity.
Newton's second law	$\vec{\mathbf{F}} = \Delta\vec{\mathbf{p}}/\Delta t$	The average force on an object equals the rate of change of the object's momentum.
conservation of momentum	$m_1\,\vec{\mathbf{v}}_1 + m_2\,\vec{\mathbf{v}}_2 = m_1\,\vec{\mathbf{v}}_1' + m_2\,\vec{\mathbf{v}}_2'$	In a collision between two or more objects the total momentum is conserved.
impulse	$\vec{\mathbf{F}}\,\Delta t$	Impulse equals the product of the force and the time during which the force acts.
impulse and momentum	$\vec{\mathbf{F}}\,\Delta t = \Delta\vec{\mathbf{p}} = m\,\vec{\mathbf{v}}' - m\,\vec{\mathbf{v}}$	An impulse equals the change in an object's momentum.
coordinate positions of the center of mass	$x_{CM} = (\Sigma m_i\, x_i)/(\Sigma m_i)$ $y_{CM} = (\Sigma m_i\, y_i)/(\Sigma m_i)$	These equations are used to determine the horizontal and/or vertical position of the center of mass.
translational velocity of the center of mass	$M\,\vec{\mathbf{v}}_{cm} = m_1\,\vec{\mathbf{v}}_1 + m_2\,\vec{\mathbf{v}}_2 + m_3\,\vec{\mathbf{v}}_3$	This equation is used to determine the translational velocity of the center of mass.

CONCEPT SUMMARY

Linear Momentum and Force

Linear momentum ($\vec{\mathbf{p}}$) is defined as the product of an object's mass (m) and its velocity ($\vec{\mathbf{v}}$). Momentum is a vector quantity and the direction of the momentum vector is the same as the velocity vector. The formula for linear momentum is

$\vec{\mathbf{p}} = m\,\vec{\mathbf{v}}$ The unit of linear momentum is kg m/s. This is NOT the same as the unit of force, which is $1\text{ N} = 1\text{ kg m/s}^2$.

Newton's second law may be written in terms of the average force required to change an object's momentum as follows:

$\vec{F} = \Delta\vec{p}/\Delta t$ where $\Delta\vec{p}$ is the change in momentum and Δt is the time interval during which the change in momentum occurs.

Conservation of Linear Momentum

In any collision between two or more objects, the vector sum of the momenta before impact equals the vector sum of the momenta after impact. The total momentum remains constant, i.e., the momentum is conserved. This principle is known as the **law of conservation of momentum**. For a collision between two objects, the law of conservation of momentum can be written as follows:

$$m_1 \vec{v}_1 + m_2 \vec{v}_2 = m_1 \vec{v}_1{}' + m_2 \vec{v}_2{}'$$

where $\vec{v}_1$ and $\vec{v}_2$ are the velocities of the objects before impact and $\vec{v}_1{}'$ and $\vec{v}_2{}'$ are the velocities of the objects after impact.

TEXTBOOK QUESTION 4. It is said that in ancient times a rich man with a bag of gold coins was frozen to death while stranded on the surface of a frozen lake. Because the ice was frictionless, he could not push himself to shore. What could he have done to save himself had he not been so miserly?

ANSWER: Assuming that the man is at rest on the ice, the initial momentum of the system is zero. If he should throw the bag of coins, he would recoil in the opposite direction. His mass is greater than that of the coins and therefore his recoil speed would be much smaller than the speed of the bag of coins. The final momentum of the coins is equal but opposite the man's momentum and the total final momentum equals zero. Thus, the law of conservation of momentum is upheld. Since the ice is frictionless, his recoil velocity would remain constant until he reached the bank. At that point he could walk around to the opposite bank and retrieve the bag of coins.

An alternate possibility would be to keep the coins in his pocket and throw his shoe or boot. The result would be the same and he would save himself and his gold.

TEXTBOOK QUESTION 10. A light body and a heavy body have the same kinetic energy. Which has the greater momentum?

ANSWER: Assume that the light body has a mass of 1.0 kg and the mass of the heavy body is 4.0 kg. If the speed of the light body is 4.0 m/s then its KE = ½ (1.0 kg)(4.0 m/s)^2 = 8.0 J. The speed of the heavy body can now be determined because it has an equal amount of KE:

$8.0\ J = ½\ (4.0\ kg)\ v^2$ and $v = 2.0$ m/s.

Momentum is the product of the mass and the velocity. The momentum of the light body is (1.0 kg)(4.0 m/s) = 4.0 kg m/s and the momentum of the heavy body is (4.0 kg)(2.0 m/s) = 8.0 kg m/s. Thus, while both objects have the same KE, the heavy object has twice as much momentum.

Impulse

An **impulse** is defined as the product of the force ($\vec{\mathbf{F}}$) acting on an object and the time (Δt) during which the force acts.

$$\text{impulse} = \vec{\mathbf{F}}\,\Delta t$$

The unit of impulse is newton seconds (N s) where 1 N s = 1 kg m/s. Impulse is a vector quantity and the direction of the impulse vector is the same as that of the force vector.

An impulse causes a change in an object's momentum. The mathematical formula that connects impulse and change in momentum is as follows:

$$\vec{\mathbf{F}}\,\Delta t = \Delta\vec{\mathbf{p}} = m\,\vec{\mathbf{v}}' - m\,\vec{\mathbf{v}}$$

In a collision between two or more objects, the force each object exerts on the other is usually very large compared to any other force acting and the time interval during which the interaction occurs is usually very short. Usually, both the magnitude of the force and the time interval remain unknown; however, the impulse can be determined if it is possible to determine the change in momentum.

TEXTBOOK QUESTION 7. Cars used to be built as rigid as possible to withstand collisions. Today, though, cars are designed to have "crumple zones" that collapse upon impact. What is the advantage of this new design?

ANSWER: During a collision an outside impulse acts on the car and changes the car's momentum. The change in momentum of the car remains the same whether there are crumple zones or not. However, crumple zones increase the time during which the force exerted by the outside object acts. The increase in time results in a smaller average force acting on the car (and the occupants). Assuming the crumple zones are properly designed, the kinetic energy is dissipated in the form of heat energy and the deformation of the metal.

As an added safety measure, the crumple zones in modern cars are designed to crumple around the passenger compartment. The desired result is that the passenger compartment will remain intact and the passengers are protected from major injury.

EXAMPLE PROBLEM 1. A 0.0600 kg tennis ball is traveling at 30.0 m/s. After being hit by the opponent's racket, the ball's velocity is 20.0 m/s in the opposite direction. Compute the a) change in the ball's momentum and b) average force exerted by the racket if the ball and racket were in contact for 0.0400 s. Hint: Assume that the ball's initial direction of motion is the positive direction.

Part a. Step 1.

Determine the change in the tennis ball's momentum.

Solution: (Section 7-3)

Velocity is a vector quantity. Therefore, if the initial velocity is +30.0 m/s, the final velocity is a -20.0 m/s.

$$\Delta\vec{\mathbf{p}} = m\,\vec{\mathbf{v}}' - m\,\vec{\mathbf{v}}$$

$$= (0.0600\ \text{kg})(-20.0\ \text{m/s}) - (0.0600\ \text{kg})(+30.0\ \text{m/s})$$

	$\vec{\Delta p}$ = -1.92 kg m/s
Part b. Step 1. Use the impulse-momentum equation to determine the net force acting on the ball.	$\vec{F} \Delta t = m \vec{v}' - m \vec{v}$ $\mathbf{F}$ (0.0400 s) = -1.92 kg m/s $\vec{F}$ = -48.0 N Note: the negative sign indicates that the direction of the force is in the direction arbitrarily defined as the negative direction, i.e., in the same direction that the ball is moving after impact.

Energy and Momentum Conservation in Collisions

In collisions between two (or more) objects, the total momentum is always conserved. However, some of the kinetic energy may be converted into other forms of energy, usually thermal energy. Problems involving two types of collisions can be solved by using the methods of this chapter: perfectly elastic collisions and completely inelastic collisions.

In **perfectly elastic collisions,** the kinetic energy as well as the linear momentum is conserved. The sum of the kinetic energies of the objects before the collision is equal to the sum of the kinetic energies of the objects after the collision. No heat energy results from the collision. Certain perfectly elastic collision problems can be solved by applying both the law of conservation of energy and the law of conservation of momentum.

In **completely inelastic collisions**, the objects stick together after impact. Most of the kinetic energy is converted into heat. This type of problem can be solved by applying the law of conservation of momentum.

EXAMPLE PROBLEM 2. As shown in the diagram, a bullet of mass 0.0500 kg traveling at 50.0 m/s is fired horizontally into a wooden block suspended from a long rope. The mass of the wooden block is 0.300 kg and it is initially at rest. The collision is completely inelastic and after impact the bullet + wooden block move together until the center of mass of the system rises a vertical distance h above its initial position. Calculate the a) velocity of the bullet + wooden block just after impact and b) vertical distance h reached by the bullet + wooden block.

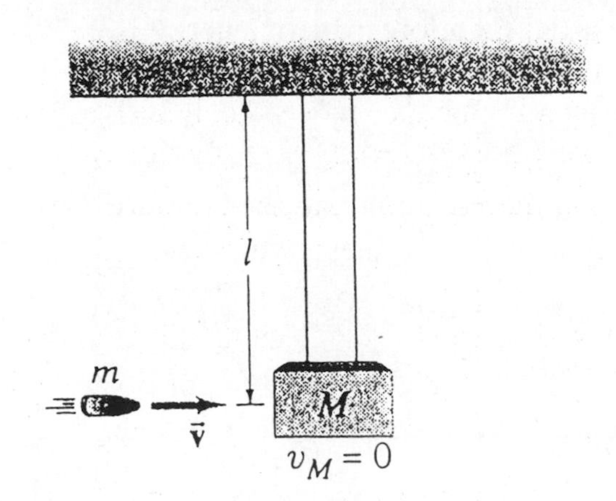

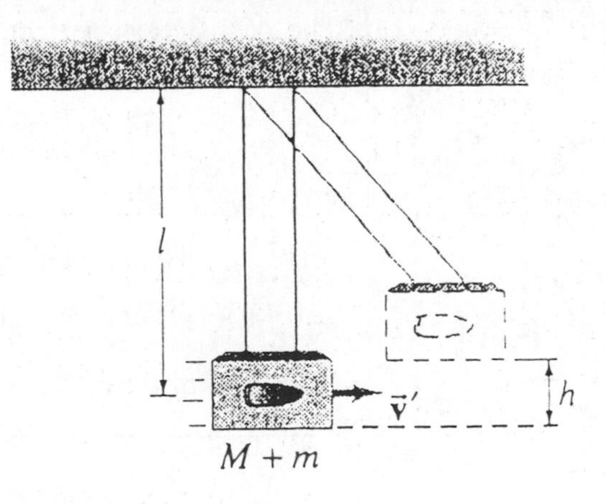

Part a. Step 1. Use the law of conservation of momentum to determine the velocity of the bullet + pendulum just after impact.	Solution: (Section 7-6) The initial momentum equals the final momentum. The collision is completely inelastic. The bullet and wooden block stick together after impact; therefore, $\vec{v}_b' = \vec{v}_w' = \vec{v}'$ $m_b \vec{v}_b + m_w \vec{v}_w = (m_b + m_w) \vec{v}'$ $(0.0500 \text{ kg})(50.0 \text{ m/s}) + (0.300 \text{ kg})(0 \text{ m/s}) = (0.0500 \text{ kg} + 0.300 \text{ kg}) \vec{v}'$ $\vec{v}' = (2.50 \text{ kg m/s})/0.350 \text{ kg})$ $\vec{v}' = 7.14 \text{ m/s}$
Part b. Step 1. The collision is completely inelastic. The objects stick together after impact. Use the law of conservation of energy to solve for the velocity just after impact.	$KE_{\text{after impact}} = PE_{\text{at top of swing}}$ $\frac{1}{2}(m_b + m_w)v'^2 = (m_b + m_w) g h$ the mass cancels out, and rearranging gives $\frac{1}{2} v'^2 = g h$ $\frac{1}{2}(7.14 \text{ m/s})^2 = (9.8 \text{ m/s}^2) h$ and $h = 2.60 \text{ m}$

EXAMPLE PROBLEM 3. A 2.0 kg object traveling at 1.0 m/s collides head-on with a 1.0 kg object initially at rest. Determine the velocity of each object after the impact if the collision is perfectly elastic.

Part a. Step 1. Write an equation for the collision using the law of conservation of momentum.	Solution: (Sections 7-4 and 7-5) $m_1 \vec{v}_1 + m_2 \vec{v}_2 = m_1 \vec{v}_1' + m_2 \vec{v}_2'$ $(2.0 \text{ kg})(1.0 \text{ m/s}) + (1.0 \text{ kg})(0 \text{ m/s}) = (2.0 \text{ kg}) \vec{v}_1' + (1.0 \text{ kg}) \vec{v}_2'$ $2 \text{ m/s} = 2 \vec{v}_1' + 1 \vec{v}_2'$ (equation 1)
Part a. Step 2. Write an equation for the collision using the law of conservation of energy.	The collision is perfectly elastic; therefore, kinetic energy is conserved. $\frac{1}{2} m_1 v_1^2 + \frac{1}{2} m_2 v_2^2 = \frac{1}{2} m_1 v'^2 + \frac{1}{2} m_2 v'^2$ Since ½ appears in each term, the equation may be simplified by dividing each term by ½. Substituting values gives $(2.0 \text{ kg})(1.0 \text{ m/s})^2 + (1.0 \text{ kg})(0 \text{ m/s})^2 = (2.0 \text{ kg}) v_1'^2 + (1.0 \text{ kg}) v_2'^2$ Upon simplifying, $2 \text{ m}^2/\text{s}^2 = 2 v_1'^2 + 1 v_2'^2$ (equation 2)

Part a. Step 3.

Application of the conservation laws has resulted in two algebraic equations with the same two unknowns. Solve for the final velocity of each object.

from equation 1: $\vec{\mathbf{v}}_2' = 2 - 2\,\vec{\mathbf{v}}_1'$

Substituting for $\vec{\mathbf{v}}_2'$ in equation 2,

$2 = 2\,v_1'^2 + (2 - 2\,v_1')^2$

$2 = 2\,v_1'^2 + 4 - 8\,v_1' + 4\,v_1'^2$

$0 = 6\,v_1'^2 - 8\,v_1' + 2$ and simplifying gives

$0 = 3\,v_1'^2 - 4\,v_1' + 1$ and factoring gives

$0 = (3\,v_1' - 1)(v_1' - 1)$

Either $0 = 3\,v_1' - 1$ or $v_1' - 1 = 0$

and either $\vec{\mathbf{v}}_1' = ⅓$ m/s or $\vec{\mathbf{v}}_1' = 1$ m/s

According to Newton's third law, after impact the velocity of object 1 must change. Therefore, the solution, $v_1' = 1$, while a valid solution, is uninteresting because it means that object 1 missed object 2, so object 2 is still at rest. The interesting answer is $v_1' = ⅓$ m/s. Substitute the value of v_1' into equation 1 and determine v_2'.

$\vec{\mathbf{v}}_2' = 2 - 2\,v_1' = 2 - 2\,(⅓)$

$\vec{\mathbf{v}}_2' = 4/3$ m/s $= 1.3$ m/s

Center of Gravity and Center of Mass

The **center of gravity** (cg) of an object is the point where the entire weight of the object can be considered to be concentrated. At that point the entire weight of the object can be balanced by a single vertical force equal to the object's weight. For example, you can test for the center of gravity of a book by attempting to balance the book with one finger. The center of gravity of a book should be very close to the center of the book. For a baseball bat, the balance point and therefore the center of gravity is displaced from the center toward the fat end of the bat.

The **center of mass,** which is abbreviated CM, is the point where all of the mass of an object can be considered to be concentrated. For almost all objects, the center of mass is at the same point in the object as the center of gravity. An extreme example to show the difference is a thin, uniform rod that extends from the surface of the Earth vertically upward. The center of mass of the rod would be at its center. However, since the value of g decreases with altitude and the rod is extremely long, the lower half of the rod would weigh more than the upper half. In this case the center of gravity would fall below the center of mass.

The following formula can be used to determine the distance from one end to the center of mass of a massless rod along which a number of point masses have been attached.

$$x_{CM} = (m_1\,x_1 + m_2\,x_2 + \ldots + m_n\,x_n)/(m_1 + m_2 + \ldots + m_n)$$

EXAMPLE PROBLEM 4. The mass of the Sun is 2.0×10^{30} kg while the mass of the Earth is 6.0×10^{24} kg. The center-to-center distance between the Earth and the Sun is 1.50×10^{11} m. Determine the distance from the center of the Sun to the center of mass of the Earth-Sun system.

	Solution: (Section 7-8)
Part a. Step 1. The center of mass is to be measured from the center of the Sun. Therefore, $x_{Sun} = 0$ m. Complete a data table listing the mass and position of each object.	$m_{Sun} = 2.0 \times 10^{30}$ kg, $x_{Sun} = 0$ m $m_{Earth} = 6.0 \times 10^{24}$ kg, $x_{Earth} = 1.50 \times 10^{11}$ m
Part a. Step 2. Use the formula for determining the position of the center of mass of objects arranged along a horizontal line.	$x_{CM} = (m_{Sun}\, x_{Sun} + m_{Earth}\, x_{Earth})/(m_{Sun} + m_{Earth})$ $m_{Sun}\, x_{Sun} = (2.0 \times 10^{30}\ \text{kg})(0.0\ \text{m}) = 0$ kg m $m_{Earth}\, x_{Earth} = (6.0 \times 10^{24}\ \text{kg})(1.50 \times 10^{11}\ \text{m}) = 9.0 \times 10^{35}$ kg m $m_{Sun} + m_{Earth} = 2.0 \times 10^{30}\ \text{kg} + 6.0 \times 10^{24}\ \text{kg} \approx 2.0 \times 10^{30}$ kg $x_{CM} = (0\ \text{kg m} + 9.0 \times 10^{35}\ \text{kg m})/(2.0 \times 10^{30}\ \text{kg})$ $x_{CM} = 4.5 \times 10^{5}\ \text{m} \approx 280$ miles

Note: the Earth-Sun system revolves around the center of mass. The inside cover of the textbook lists the radius of the Sun as 6.96×10^{8} m (approximately 433,000 miles). Therefore, the center of mass of the Earth-Sun system is very close (approximately 280 miles) from the center of the Sun. For all practical purposes, the Earth revolves about the center of the Sun.

Center of Mass and Translational Motion

The term **translational motion** is used in situations where the motion of all points of a moving object have at any instant the same velocity and direction of motion. This type of motion, as well as the concept of center of mass, has been implied in solving problems in previous chapters of the textbook. For example, the free-body diagrams used in chapter 3 assumed that the mass of an object could be considered to be concentrated at a point with its weight acting at that point.

The concept of center of mass is also useful when discussing the motion of a group of particles. For example, as stated in the textbook, "The total linear momentum of a system of particles is equal to the product of the total mass M and the velocity of the center of mass of the system. Or, the linear momentum of an extended body is the product of the body's mass and the velocity of its CM." Therefore,

$$M\,\vec{v}_{cm} = m_1\,\vec{v}_1 + m_2\,\vec{v}_2 + m_3\,\vec{v}_3 + \ldots + m_n\,\vec{v}_n$$

Also, Newton's second law can be written

$$\Sigma \mathbf{F} = M\,\vec{a}_{CM}$$

And as stated in the textbook, "The sum of all of the forces acting on the system is equal to the total mass of the system times the acceleration of the center of mass."

PROBLEM SOLVING SKILLS

For problems involving impulse-change of momentum:

1. Draw an accurate diagram locating all of the forces acting on the system.
2. If a net external force acts on the object(s), then the momentum of the system will change. Determine the magnitude and direction of the net force.
3. Apply the impulse-momentum equation taking note that force and velocity are vectors and that direction of the vector plays an important part in the solution.

For problems involving no external force acting on the system:

1. Use the law of conservation of momentum to solve the problem. Take note that momentum is a vector quantity and must be considered in the solution.

For problems involving graphical integration:

1. Determine the sum of the partial and complete blocks that lie under the curve.
2. Determine the magnitude of the impulse represented by one block.
3. Determine the magnitude of the total impulse by multiplying the impulse represented by one block by the total number of blocks found in Step 2.
4. Use the impulse-momentum equation and solve the problem.

For problems involving perfectly elastic and completely inelastic collisions:

1. Determine which type of collision is described in the problem.
2. Use the law of conservation of momentum to solve problems where the collision is inelastic.
3. If the collision is perfectly elastic use both conservation of momentum and conservation of mechanical energy. Each law produces an algebraic equation with two unknowns. The final velocity of each object can be determined by applying standard algebraic techniques.

SOLUTIONS TO SELECTED TEXTBOOK PROBLEMS

TEXTBOOK PROBLEM 4. A child in a boat throws a 6.40 kg package out horizontally with a speed of 10.0 m/s, Fig. 7-31. Calculate the velocity of the boat immediately after, assuming it was initially at rest. The mass of the child is 26.0 kg and that of the boat is 45.0 kg. Ignore water resistance.

Part a. Step 1. Complete a data table including information both given and implied.	Solution: (Section 7-3) $m_1 = 6.40$ kg, $v_1 = 0$ m/s, $v_1' = +10.0$ m/s $m_2 = 26.0$ kg, $v_2 = 0$ m/s $v_2' = ?$ $m_3 = 45.0$ kg, $v_3 = 0$ m/s $v_3' = ?$ but the child and the boat recoil together; therefore, $\vec{v}_2' = \vec{v}_3' = \vec{v}' = ?$
Part a. Step 2. Apply the law of conservation of momentum.	$m_1 \vec{v}_1 + m_2 \vec{v}_2 + m_3 \vec{v}_3 = m_1 \vec{v}_1' + m_2 \vec{v}_2' + m_3 \vec{v}_3'$ $(6.40 \text{ kg})(0 \text{ m/s}) + (26.0 \text{ kg})(0 \text{ m/s}) + (45.0 \text{ kg})(0 \text{ m/s}) =$ $(6.40 \text{ kg})(10.0 \text{ m/s}) + [(26.0 \text{ kg}) + (45.0 \text{ kg})] \vec{v}'$

$0 = 64.0 \text{ kg m/s} + (71.0 \text{ kg}) \vec{v}'$

$\vec{v}' = - 0.901 \text{ m/s}$

Note: the negative sign indicates that the child + boat recoil in the direction opposite from the package.

TEXTBOOK PROBLEM 10. A 3800 kg open railroad car coasts along with a constant speed of 8.60 m/s along a level track. Snow begins to fall vertically and fills the car at rate of 3.50 kg/min. Ignoring friction with the tracks, what is the speed of the car after 90 min?

Part a. Step 1. Complete a data table listing information both given and implied.	Solution: (Section 7-2) $m_1 = 3800 \text{ kg}$ $m_{2initial} = 0 \text{ kg}$ $v_1 = 8.60 \text{ m/s}$ $m_{2final} = (3.50 \text{ kg/min})(90 \text{ min}) = 315 \text{ kg}$ $v' = ?$ Note: the falling snow has no horizontal component of motion; therefore, $v_2 = 0$ m/s.
Part a. Step 2. Use the law of conservation of momentum to solve for the speed of the car after 90 minutes.	$m_1 \vec{v}_1 + m_2 \vec{v}_2 = (m_1 + m'_2) \vec{v}'$ $(3800 \text{ kg})(8.60 \text{ m/s}) + (0.0 \text{ kg})(0 \text{ m/s}) = [(3800 \text{ kg}) + (3.50 \text{ kg/min})(90 \text{ min})] \vec{v}'$ $32680 \text{ kg m/s} = (4115 \text{ kg}) \vec{v}'$ $\vec{v}' = 7.94 \text{ m/s}$

TEXTBOOK PROBLEM 12. A 23 g bullet traveling at 230 m/s penetrates a 2.0 kg block of wood and emerges cleanly at 170 m/s. If the block is stationary on a frictionless surface when hit, how fast does it move after the bullet emerges?

Part a. Step 1. Complete a data table listing information both given and implied.	Solution: (Section 7-2) $m_1 = 23 \text{ g} = 0.023 \text{ kg}$ $m_2 = 2.0 \text{ kg}$ $\vec{v}_1 = 230 \text{ m/s}$ $\vec{v}_2 = 0 \text{ m/s}$ $\vec{v}_1' = 170 \text{ m/s}$ $\vec{v}_2' = ?$
Part a. Step 2. Use the law of conservation of momentum to solve for the final speed of the block of wood.	$m_1 \vec{v}_1 + m_2 \vec{v}_2 = m_1 \vec{v}_1' + m_2 \vec{v}_2'$ $(0.023 \text{ kg})(230 \text{ m/s}) + (2.0 \text{ kg})(0 \text{ m/s}) = (0.023 \text{ kg})(170 \text{ m/s}) + (2.0 \text{ kg}) \vec{v}_2'$ $5.29 \text{ kg m/s} = (3.91 \text{kg m/s}) + (2.0 \text{ kg}) \vec{v}_2'$ $\vec{v}_2' = 0.69 \text{ m/s}$

TEXTBOOK PROBLEM 15. A golf ball of mass 0.045 kg is hit off the tee at a speed of 45 m/s. The golf club was in contact with the ball for 3.5×10^{-3} s. Find (a) the impulse imparted to the golf ball, and (b) the average force exerted on the ball by the golf club.

Part a. Step 1. Complete a data table listing information both given and implied.	Solution: (Section 7-3) $F = ?$ $\quad \Delta t = 3.5 \times 10^{-3}$ s $\quad m = 0.045$ kg $v = 0$ m/s $\quad v' = 45$ m/s
Part a. Step 2. Use the impulse-momentum equation to determine the impulse imparted to the golf ball.	impulse = $\vec{\mathbf{F}}\,\Delta t$ but $\vec{\mathbf{F}}\,\Delta t = m\,\vec{\mathbf{v}}' - m\,\vec{\mathbf{v}}$ impulse = $m\,\vec{\mathbf{v}}' - m\,\vec{\mathbf{v}}$ impulse = (0.045 kg)(45 m/s) - (0.0450 kg)(0.0 m/s) impulse = 2.0 N s
Part b. Step 1. Use the impulse-momentum equation to determine the average force on the ball.	$\vec{\mathbf{F}}\,\Delta t = m\,\vec{\mathbf{v}}' - m\,\vec{\mathbf{v}}$ $\vec{\mathbf{F}}\,(3.5 \times 10^{-3}$ s) = (0.045 kg)(45 m/s) - (0.0450 kg)(0.0 m/s) $\vec{\mathbf{F}} = 5.8 \times 10^{2}$ N

TEXTBOOK PROBLEM 23. A 0.450 kg ice puck, moving east with a speed of 3.00 m/s, has a head-on collision with a 0.900 kg puck initially at rest. Assuming a perfectly elastic collision, what will be the speed and direction of each object after the collision?

Part a. Step 1. Complete a data table listing information both given and implied.	Solution: (Section 7-4 and 7-5) $m_1 = 0.450$ kg $\quad m_2 = 0.900$ kg $\quad \vec{\mathbf{v}}_1 = 3.00$ m/s $\vec{\mathbf{v}}_2 = 0$ m/s $\quad \vec{\mathbf{v}}_1{}' = ?$ $\quad \vec{\mathbf{v}}_2{}' = ?$
Part a. Step 2. Write an equation for the collision using the law of conservation of momentum.	$m_1\,\vec{\mathbf{v}}_1 + m_2\,\vec{\mathbf{v}}_2 = m_1\,\vec{\mathbf{v}}_1{}' + m_2\,\vec{\mathbf{v}}_2{}'$ (0.450 kg)(3.00 m/s) + (0.900 kg)(0 m/s) = (0.450 kg) $\vec{\mathbf{v}}_1{}'$ + (0.900 kg) $\vec{\mathbf{v}}_2{}'$ Dividing each term by 0.450 kg, and then simplifying gives 3 m/s = $1\,\vec{\mathbf{v}}_1{}' + 2\,\vec{\mathbf{v}}_2{}'$ (equation 1)
Part a. Step 3. Write an equation for the collision using the law of conservation of energy.	The collision is perfectly elastic; therefore, kinetic energy is conserved. $\frac{1}{2} m_1 v_1{}^2 + \frac{1}{2} m_2 v_2{}^2 = \frac{1}{2} m_1 v'^2 + \frac{1}{2} m_2 v'^2$ Since ½ appears in each term, the equation may be simplified by dividing each term by ½. Substituting values and again dividing each term by 0.450 kg, and then simplifying gives

	$9\ m^2/s^2 = 1\ v_1'^2 + 2\ v_2'^2$ (equation 2)
Part a. Step 4. Application of the conservation laws has resulted in two algebraic equations with the same two unknowns. Solve for the final velocity of each object.	From equation 1: $\vec{v}_1' = 3 - 2\ \vec{v}_2'$ Substituting for $\vec{v}_1'$ in equation 2, $9 = (3 - 2\ v_2')^2 + 2\ v_2'^2$ $9 = 4\ v_2'^2 - 12\ v_2' + 9 + 2\ v_2'^2$ $0 = 6\ v_2'^2 - 12\ v_2'$ and dividing each term by 6 gives $0 = v_2'^2 - 2\ v_2'$ factor the equation and solve for v_2' $0 = (v_2')(v_2' - 2)$ Either $0 = v_2'$ or $0 = v_2' - 2$ and either $\vec{v}_2' = 0$ m/s or $\vec{v}_2' = 2.00$ m/s (east) $v_2' = 0$ indicates that object 1 missed object 2, so object 2 is still at rest. According to Newton's third law, after impact the velocity of object 2 must change. Since object 1 struck object 2, then the correct answer is $v_2' = 2.00$ m/s (east). Substitute the value of $\vec{v}_2'$ into equation 1 and determine $\vec{v}_1'$. $3\ m/s = 1\ \vec{v}_1' + 2\ \vec{v}_2'$ (equation 1) $3 = 1\ \vec{v}_1' + 2\ (2\ m/s)$ $\vec{v}_1' = -1.00$ m/s Note: the negative sign indicates that object 1 rebounds and travels west after impact.

TEXTBOOK PROBLEM 48. The CM of an empty 1050 kg car is 2.50 m behind the front of the car. How far from the front of the car will the CM be when two people sit in the front seat 2.80 m from the front of the car, and three people sit in the back seat 3.90 m from the front? Assume that each person has a mass of 70.0 kg.

Part a. Step 1. Complete a data table listing the mass and distance of each object from the front of the car.	Solution: (Section 7-8) Note: let the center of mass (CM) be measured from the front of the car. $m_{car} = 1050$ kg $x_{car} = 2.50$ m $m_{people\ in\ front} = (2)(70.0\ kg) = 140$ kg $x_{people\ in\ front} = 2.80$ m $m_{people\ in\ back} = (3)(70.0\ kg) = 210$ kg $x_{people\ in\ back} = 3.90$ m

Part a. Step 2. Use the formula for determining the position of the center of mass of objects placed along a horizontal line.	$x_{CM} = [m_{car}\, x_{car} + m_{people\ in\ front}\, x_{people\ in\ front} + m_{people\ in\ back}\, x_{people\ in\ back}]/[m_{total}]$ $m_{car}\, x_{car} = (1050\ kg)(2.50\ m) = 2625\ kg\ m$ $m_{people\ in\ front}\, x_{people\ in\ front} = (140\ kg)(2.80\ m) = 392\ kg\ m$ $m_{people\ in\ back}\, x_{people\ in\ back} = (210\ kg)(3.90\ m) = 819\ kg\ m$ $m_{total} = 1050\ kg + 140\ kg + 210\ kg = 1400\ kg$ $x_{CM} = [2625\ kg\ m + 392\ kg\ m + 819\ kg\ m]/[1400\ kg]$ $x_{CM} = 2.74\ m$

CHAPTER 8

ROTATIONAL MOTION

OBJECTIVES

After studying the material of this chapter, the student should be able to:

- convert angular quantities from revolutions or degrees to radians and vice versa.
- write the Greek symbols used to represent angular displacement, angular velocity, and angular acceleration.
- state the meaning of the symbols used in the kinematics equations for uniformly accelerated angular motion.
- write from memory the equations used to describe uniformly accelerated angular motion.
- complete a data table using information both given and implied in word problems. Use the completed data table to solve word problems related to angular kinematics.
- distinguish between inertia and moment of inertia. Write from memory the formulas for the moment of inertia of selected objects and calculate the moment of inertia of these objects.
- explain the meaning of the radius of gyration. Use the radius of gyration to solve for an object's moment of inertia.
- distinguish between linear momentum and angular momentum. State and apply the law of conservation of angular momentum to solve word problems.
- calculate the lever arm distance and determine the magnitude and direction of the torque vector if the magnitude and direction of the net force are given.
- draw a free body diagram for each object in a system. Locate the forces acting on each object. Use $F = m\ a$ and $\tau = I\ \alpha$ to solve for the linear or angular acceleration of each object.
- apply the law of conservation of angular momentum to a system where no net external torque acts. Determine the change in angular velocity of a system where the moment of inertia of the objects that make up the system changes.
- distinguish between translational kinetic energy and rotational kinetic energy. Apply the law of conservation of energy to solve problems that involve rotational as well as translational kinetic energy.

KEY TERMS AND PHRASES

radian is a dimensionless quantity and is the Système International (SI) unit for angular measurement. 1 radian $\approx 57.3^\circ$ while 1 revolution = $2\ \pi$ radians = 360°.

angular displacement (θ) is the angle through which a point on a rotating object moves during a time interval. The angular displacement is measured in radians, degrees, or revolutions.

angular velocity (ω) is measured in radians per second. The average angular velocity is measured by dividing the angular displacement by the time required to travel through the displacement.

angular acceleration (α) is measured in radians per second per second (rad/s^2). The average angular acceleration is defined as the rate of change of the angular velocity in time.

torque (τ) is the measure of the effectiveness of a force in producing rotation of an object about an axis.

moment of inertia or rotational inertia (I) is the measure of the tendency of an object to resist any change in its state of rotation.

radius of gyration is the distance from the axis of rotation to a point where all of the object's mass could be concentrated. The moment of inertia of an irregularly shaped object is often determined by using the object's radius of gyration.

rotational kinetic energy is energy due to an object's rotational motion. The rotational kinetic energy is related to the object's moment of inertia and the square of its angular velocity.

angular momentum (L) is a quantity that is found from the product of an object's moment of inertia (I) and angular velocity (ω).

law of conservation of angular momentum states that in the absence of a net torque acting on an object, the object's angular momentum must remain constant in both magnitude and direction, i.e., $I\ \omega$ = constant.

SUMMARY OF MATHEMATICAL FORMULAS

angular displacement	$\theta = \ell/r$ or $\ell = r\,\theta$	Angular displacement is directly proportional to the length of arc (ℓ) and inversely related to the radius (r).
angular velocity	$\omega = v/r$ or $v = r\,\omega$	Angular velocity (ω) is directly proportional to the tangential velocity (v) and inversely related to the radius (r).
angular acceleration	$\alpha = a_T/r$ or $a_T = r\,\alpha$.	Angular acceleration (α) is directly proportional to the tangential acceleration (a_T) and inversely proportional to the radius (r).
centripetal acceleration	$a_c = r\,\omega^2$	Centripetal acceleration equals the product of the radius and the square of the angular velocity.

equations for uniformly accelerated rotational motion	$\omega = \omega_o + \alpha t$	Angular speed is related to initial angular speed, angular acceleration, and time.
	$\theta = \omega_o t + \frac{1}{2} \alpha t^2$	Angular displacement is related to initial angular velocity, angular acceleration, and time.
	$\omega^2 - \omega_o^2 = 2 \alpha \theta$	Angular velocity is related to initial angular velocity, angular acceleration and angular displacement.
	$\bar{\omega} = (\omega + \omega_o)/2$	Average angular velocity is related to the initial angular velocity and the final angular velocity.
	$\theta = \bar{\omega} t$	Angular displacement is related to the average angular velocity and time.
torque	$\tau = \ell_\perp F = \ell F \sin \theta$	Torque is related to the length of the lever arm and the magnitude of the applied force.
torque	$\tau = I \alpha$	Torque is related to the moment of inertia and the angular acceleration.
moment of inertia Note: M represents the object's mass, R represents the object's radius, and L represents the object's length.	$I = M R^2$	Moment of inertia of a thin ring of radius R
	$I = \frac{1}{2} M R^2$	Moment of inertia of a solid cylinder of radius R
	$I = (2/5) M R^2$	Moment of inertia of a solid sphere of radius R
	$I = (1/12) M L^2$	Moment of inertia of a long rod of length L (axis through center and perpendicular to rod)
	$I = \frac{1}{3} M L^2$	Moment of inertia of a long rod of length L (axis through end and perpendicular to rod)
moment of inertia	$I = M k^2$	Moment of inertia written in terms of the object's mass and radius of gyration (k)
rotational kinetic energy	$KE = \frac{1}{2} I \omega^2$	Rotational kinetic energy is related to an object's moment of inertia and the square of its angular velocity.

angular momentum	$\vec{L} = I\vec{\omega}$	Angular momentum equals the product of an object's moment of inertia and angular velocity.
law of conservation of angular momentum	$I\vec{\omega}$ = constant	In the absence of a net torque acting on an object, the object's angular momentum must remain constant in both magnitude and direction.

CONCEPT SUMMARY

Circular Motion in Terms of Angular Quantities

Angular displacement (θ) can be measured in degrees, revolutions or radians. 1 **radian** is the angle subtended at the center of a circle by a length of arc (ℓ) equal to the radius of the circle (r), i.e., $\theta = \ell/r$. Since ℓ and r are both units of length, the radian is a quantity which is dimensionless. The radian is the Système International (SI) unit for angular measurement.
1 radian $\approx$ 57.3° while 1 revolution = 2 π radians = 360°.

Angular velocity (ω) is measured in radians per second. The average angular velocity is measured by dividing the angular displacement by the time required to travel through the displacement, $\overline{\omega} = \theta/t$. If the time interval is small, $\Delta t \to 0$, then $\omega = \Delta\theta/\Delta t$ defines the instantaneous angular velocity.

Angular acceleration (α) is measured in radians per second per second (rad/s^2). The average angular acceleration is defined as the rate of change of the angular velocity in time. If the time interval is small, $\Delta t \to 0$ and $\alpha = \Delta\omega/\Delta t$ defines the instantaneous angular acceleration.

As an object travels in a circle, the motion of a point a distance r from the center of the circle can be described in terms of tangential quantities. Since $\theta = \ell/r$, the displacement of the object from its initial position can be found by the formula $\ell = r\,\theta$. The **tangential velocity** is given by the formula $v = r\,\omega$ while the **tangential acceleration** $a_T = r\,\alpha$.

The total linear acceleration of a particle traveling in a circle is $\vec{\mathbf{a}} = \vec{\mathbf{a}}_T + \vec{\mathbf{a}}_R$. The centripetal acceleration is given by $a_R = v^2/r$, but $v = r\,\omega$, thus $a_R = \omega^2 r$.

The relationship between the frequency of rotation (f) and the angular velocity (ω) is given by the equation: $\omega = 2\,\pi\,f$. If the frequency of rotation is 1.0 revolution per second, then the angular frequency $\omega = 2\,\pi$ (1.0 rev/s) = 6.28 radians per second.

Kinematics Equations for Uniformly Accelerated Rotational Motion

The equations used for uniformly accelerated rotational motion are analogous to the equations used for uniformly accelerated linear motion. θ is analogous to $x - x_o$, ω to v, and α to a.

$v = v_o + at$ $\qquad\qquad$ $\omega = \omega_o + \alpha\,t$

$x - x_o = v_o t + \frac{1}{2} a t^2$ $\theta = \omega_o t + \frac{1}{2} \alpha t^2$

$v^2 - v_o^2 = 2 a (x - x_o)$ $\omega^2 - \omega_o^2 = 2 \alpha \theta$

$\bar{v} = (v + v_o)/2$ $\bar{\omega} = (\omega + \omega_o)/2$

$x - x_o = \bar{v}\ t$ $\theta = \bar{\omega}\ t$

In solving problems using the above formulas for rotational motion, it is important to be consistent in the use of units. It is suggested that the student always convert from revolutions or degrees to radians. Also, time is to be expressed in seconds.

EXAMPLE PROBLEM 1. A wheel 1.00 m in radius is rotating at 100 rad/s. The wheel accelerates uniformly from 100 rad/s to 500 rad/s in 100 revolutions. Determine the a) rate of angular acceleration, b) time required for the wheel to accelerate to 500 radians per second, c) tangential acceleration of a point on the rim as the wheel is accelerating, and d) radial acceleration of a point on the rim before the acceleration begins.

Part a. Step 1. Complete a data table based the information given in the problem.	Solution: (Sections 8-1 and 8-2) Since the angular velocities are given in radians per second, it is necessary only to convert the angular displacement to radians. $\theta = (100\ \text{rev})(2\pi\ \text{rad/1 rev}) = 628\ \text{rad}$ $\omega_o = 100\ \text{rad/s}$ $\alpha = ?$ $\omega = 500\ \text{rad/s}$ $t = ?$
Part a. Step 2. Solve for the angular acceleration.	$2\,\alpha\,\theta = \omega^2 - \omega_o^2$ $2\,\alpha\,(628\ \text{rad}) = (500\ \text{rad/s})^2 - (100\ \text{rad/s})^2$ $\alpha = 191\ \text{rad/s}^2$
Part b. Step 1. Determine the time required for the wheel to accelerate to 500 rad/s.	The angular acceleration is now known and can be added to the data table. Substitute the values into the appropriate equation and solve for the time. $\omega = \omega_o + \alpha\,t$ $500\ \text{rad/s} = 100\ \text{rad/s} + (191\ \text{rad/s}^2)\,t$ $t = 2.09\ \text{s}$
Part c. Step 1. Determine the tangential acceleration.	$a_T = r\,\alpha = (1.00\ \text{m})(191\ \text{rad/s}^2)$ $a_T = 191\ \text{m/s}^2$

Part d. Step 1.

Determine the magnitude and direction of the centripetal acceleration a point on the rim.

The centripetal acceleration is directed radially inward toward the center of the circle and is related to the angular velocity and radius of the circle as follows:

$a_R = r\,\omega_o^2 = (1.00\text{ m})(100\text{ rad/s})^2$

$a_R = 1.00 \times 10^4\text{ m/s}^2$

Torque

Torque (τ) is the measure of the effectiveness of a force in producing rotation of an object about an axis. It is measured by the product of the force and the perpendicular distance from the axis of rotation to the line along which the force acts. This perpendicular distance is often referred to as the **lever arm distance** ($\ell_\perp$).

$\tau = \ell_\perp F = \ell F \sin\theta$ The SI unit of torque is the meters newtons (m N).

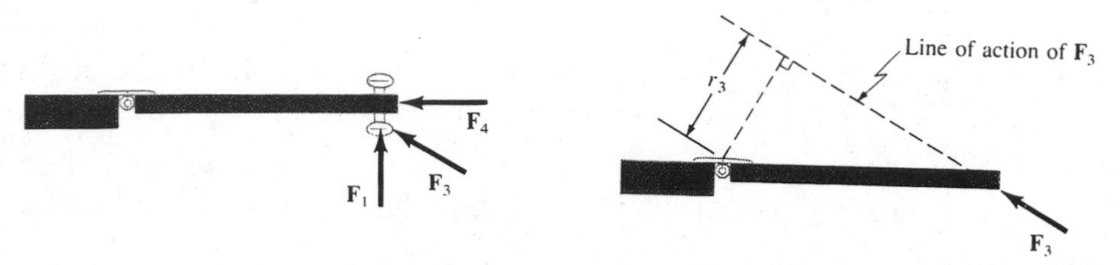

For example, from everyday experience the student realizes that the most effective manner in which to open a door is to push on the side of the door opposite from the hinge point (axis of rotation) and to apply the force perpendicular to the plane of the door. The student is applying an unbalanced torque to the door which causes it to rotate.

TEXTBOOK QUESTION 4. Can a small force ever exert a greater torque than a large force? Explain.

ANSWER: Yes, a small force can exert a greater torque than a large force. Torque equals the product of the force and the lever arm distance. A force of 5.0 N with a lever arm distance of 3.0 m produces a torque of 15 m N. A 10 N force with a lever arm distance of 1.0 m produces a torque of 10.0 m N. Therefore, if the lever arm distance is large enough, the torque exerted by the small force will be greater than that exerted by a large force with a small lever arm distance.

Rotational Dynamics; Torque and Rotational Inertia

When an unbalanced torque acts on an object, it tends to change an object's state of rotation, i.e., it produces an angular acceleration or deceleration. However, the magnitude of the angular acceleration or deceleration depends on the object's **moment of inertia** (I) as well as the magnitude of the torque (τ).

$\tau = I\,\alpha$ and $\alpha = \tau/I$

Moment of Inertia or Rotational Inertia

Just as objects tend to resist any change in translational motion (straight line motion), an object tends to resist any change in its rotational motion. The tendency to resist any change in translational motion is referred to as inertia and is measured by measuring an object's mass in kg. **Moment of inertia** or **rotational inertia** is the measure of the tendency of an object to resist any change in its state of rotation.

The moment of inertia is determined by calculating the sum of the moments of inertia of the particles that make up the object. It is determined not only by the mass of the object but also by the distribution of the mass about the axis of rotation. In general, it is necessary to use integral calculus to determine an object's moment of inertia.

The object's moment of inertia is determined by applying the formula $I = \Sigma\, m\, r^2$, where I is the symbol for moment of inertia in kg m^2, Σ is the Greek letter sigma and means "the sum of," m is the mass of a particle of the object, and r is the distance from the axis of rotation to a particular particle.

The following is a list of the moments of inertia of certain solids of uniform composition. Except for the last object, the axis of rotation in each case is through the center of the object. It is worthwhile to commit these formulas to memory.

Thin Ring of Radius R	$I = M\, R^2$
Solid Cylinder of Radius R	$I = ½\, M\, R^2$
Solid Sphere of Radius R	$I = (2/5)\, M\, R^2$
Long Rod of Length L (axis through center and perpendicular to rod)	$I = (1/12)\, M\, L^2$
Long Rod of Length L (axis through end and perpendicular to rod)	$I = ⅓\, M\, L^2$

Note: M is the object's total mass and R is the object's radius.

The moment of inertia of an object may also be written in terms of the object's **radius of gyration** (k) as $I = M\, k^2$. The radius of gyration is the distance from the axis of rotation to a point where all of the object's mass could be concentrated. The moment of inertia has the same magnitude as that determined by using the more general formula $I = \Sigma m r^2$. It is convenient to determine the moment of inertia of irregularly shaped objects by using the object's radius of gyration.

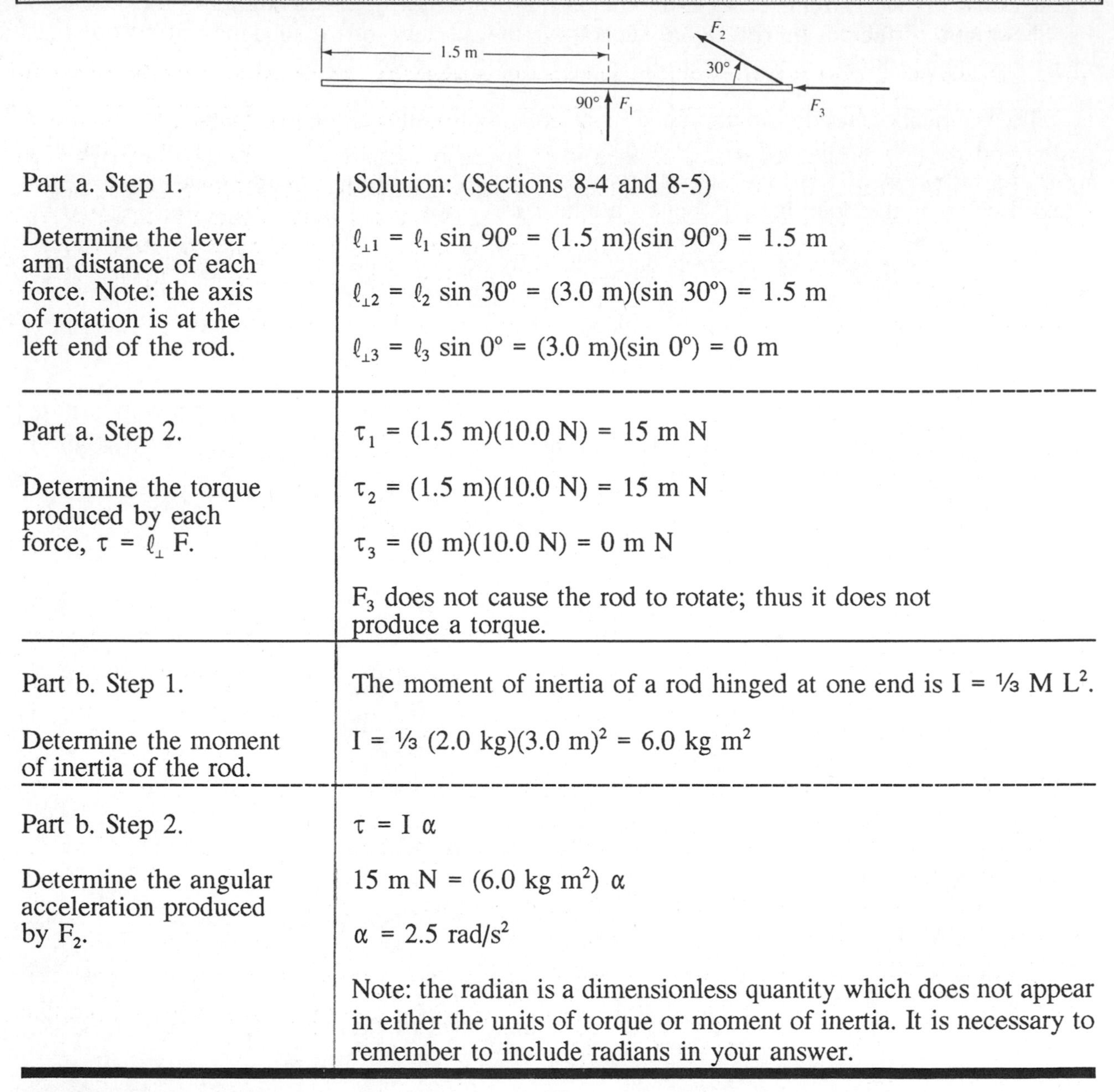

EXAMPLE PROBLEM 2. A 3.0 m long rod of mass 2.0 kg is acted upon by three forces each of magnitude 10.0 newtons as shown in the diagram. a) Determine the torque produced about the left end of the rod if each force acts individually. b) Determine the angular acceleration which results from the torque produced by F_2.

Part a. Step 1. Determine the lever arm distance of each force. Note: the axis of rotation is at the left end of the rod.	Solution: (Sections 8-4 and 8-5) $\ell_{\perp 1} = \ell_1 \sin 90° = (1.5\text{ m})(\sin 90°) = 1.5\text{ m}$ $\ell_{\perp 2} = \ell_2 \sin 30° = (3.0\text{ m})(\sin 30°) = 1.5\text{ m}$ $\ell_{\perp 3} = \ell_3 \sin 0° = (3.0\text{ m})(\sin 0°) = 0\text{ m}$
Part a. Step 2. Determine the torque produced by each force, $\tau = \ell_\perp F$.	$\tau_1 = (1.5\text{ m})(10.0\text{ N}) = 15\text{ m N}$ $\tau_2 = (1.5\text{ m})(10.0\text{ N}) = 15\text{ m N}$ $\tau_3 = (0\text{ m})(10.0\text{ N}) = 0\text{ m N}$ F_3 does not cause the rod to rotate; thus it does not produce a torque.
Part b. Step 1. Determine the moment of inertia of the rod.	The moment of inertia of a rod hinged at one end is $I = ⅓ M L^2$. $I = ⅓ (2.0\text{ kg})(3.0\text{ m})^2 = 6.0\text{ kg m}^2$
Part b. Step 2. Determine the angular acceleration produced by F_2.	$\tau = I \alpha$ $15\text{ m N} = (6.0\text{ kg m}^2)\ \alpha$ $\alpha = 2.5\text{ rad/s}^2$ Note: the radian is a dimensionless quantity which does not appear in either the units of torque or moment of inertia. It is necessary to remember to include radians in your answer.

Rotational Kinetic Energy

A rotating object has the ability to do work and therefore has energy. This energy is in the form of **rotational kinetic energy** and is given by the formula

$$KE = \tfrac{1}{2} I \omega^2$$

The total kinetic energy of an object that has both translational as well as rotational kinetic energy, for example, a wheel on a moving car, can be expressed as follows:

$KE = \frac{1}{2} M v_{CM}^2 + \frac{1}{2} I_{CM} \omega^2$

v_{CM} is the linear velocity of the center of mass and I_{CM} is the object's moment of inertia about an axis through the object's center of mass.

EXAMPLE PROBLEM 3. One end of a string is attached to a student's finger while the other end is wrapped around a solid cylinder. The cylinder is released from rest while the student holds her finger steady. The string unwinds as the cylinder rotates and accelerates downward. The mass of the cylinder is 0.20 kg and its radius is 0.030 m. Use rotational dynamics to determine the a) rate of acceleration of the object and the tension in the string. Determine the cylinder's speed when it is 0.50 m below its initial position using the b) kinematics equations and c) law of conservation of energy.

Part a. Step 1. Draw a free body diagram and locate the forces acting on the cylinder.	Solution: (Sections 8-5, 8-6, and 8-7) 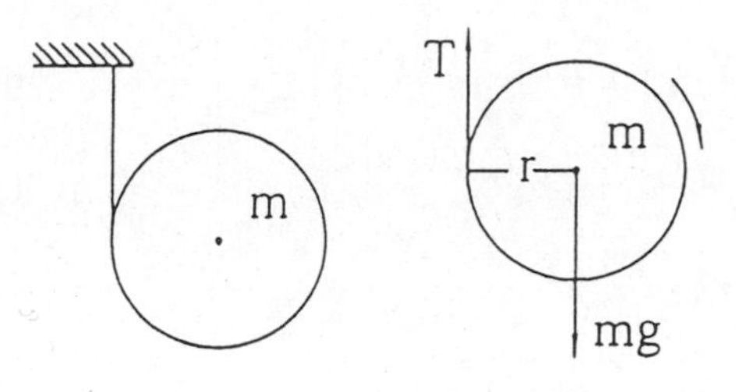
Part a. Step 2. Torque produced by tension in the string causes the cylinder to rotate. Use the torque equations and rotational dynamics to derive an equation for the tension in the string in terms of the cylinder's. rate of acceleration.	$\tau = r F \sin 90°$ but $F = T$ where T = tension in the string Also, $\tau = I \alpha$ where $I = \frac{1}{2} m r^2$ Therefore, $r T \sin 90° = \frac{1}{2} m r^2 \alpha$ and $T = \frac{1}{2} m r \alpha$ but $r \alpha = a$ Therefore, $T = \frac{1}{2} m a$ (equation 1)
Part a. Step 3. Use Newton's second law to write an equation for the rate of acceleration.	The cylinder is accelerating downward; therefore, mg is greater than the tension in the string. net force = m a $m g - T = m a$ (equation 2)
Part a. Step 4. Solve for the rate of acceleration.	From Step 2, $T = \frac{1}{2} m a$ and substituting into equation 2 gives $m g - \frac{1}{2} m a = m a$ $m g = m a + \frac{1}{2} m a$

	$m\ g = 3\ m\ a/2$ The mass cancels and $a = \frac{2}{3}\ g = \frac{2}{3}(9.8\ m/s^2) = 6.5\ m/s^2$
Part a. Step 5. Solve for the tension.	$T = \frac{1}{2}\ m\ a = \frac{1}{2}\ (0.20\ kg)(6.5\ m/s^2)$ $T = 0.65\ N$
Part b. Step 1. Use the kinematics equations to solve for the velocity of the cylinder after it travels 0.50 m.	$2\ a\ (x - x_o) = v^2 - v_o^2$ $2\ (6.5\ m/s^2)(0.50\ m) = v^2 - (0\ m/s)^2$ $v^2 = 6.5\ m^2/s^2$ $v = 2.6\ m/s$
Part c. Step 1. Solve for the velocity using the law of conservation of energy. Note: the initial KE = 0 and assume that the final PE = 0. Also, for a solid cylinder $I = \frac{1}{2}\ m\ r^2$.	$KE_{initial} + PE_{initial} = KE_{final} + PE_{final}$ $0 + m\ g\ y = \frac{1}{2}\ m\ v^2 + \frac{1}{2}\ I\ \omega^2 + 0$ but, $I = \frac{1}{2}\ m\ r^2$ and $v = r\ \omega$ Therefore, $\frac{1}{2}\ I\ \omega^2 = \frac{1}{2}\ (\frac{1}{2}\ m\ r^2)\ \omega^2 = \frac{1}{4}\ m\ v^2$ $mgy = \frac{1}{2}\ m\ v^2 + \frac{1}{4}\ m\ v^2$ $m\ g\ y = \frac{3}{4}\ m\ v^2$ m cancels, rearranging $v^2 = 4/3\ g\ y = 4/3\ (9.8\ m/s^2)(0.50\ m) = 6.5\ m^2/s^2$ $v = 2.6\ m/s$

Angular Momentum

Angular momentum ($\vec{L}$) is a quantity that is found from the product of an object's moment of inertia and angular velocity.

$$\vec{L} = I\ \vec{\omega}$$

The units of angular momentum are kg m^2/s and angular momentum is a vector quantity. The direction of the vector is along the axis of rotation and is found by the right-hand rule. The right-hand rule states that when the fingers of the right hand curl in the direction in which the object is rotating, the thumb of the right hand points in the direction of the angular momentum vector.

TEXTBOOK QUESTION 20. In what direction is the Earth's angular velocity as it rotates daily about its axis?

ANSWER: The right-hand rule is used to answer this question. The fingers of the right hand point toward the east, which is the direction of Earth's rotation. The thumb of the right hand points toward the north. Therefore, the angular velocity vector points toward the north.

Conservation of Angular Momentum

A torque tends to change an object's angular momentum and the relationship between torque and angular momentum is described by the following equation:

$$\vec{\tau} = \Delta\vec{L}/\Delta t$$

The **law of conservation of angular momentum** states that in the absence of a net torque acting on an object, the object's angular momentum must remain constant in both magnitude and direction, i.e., $I\omega = \text{constant}$ or $I_{initial}\ \omega_{initial} = I_{final}\ \omega_{final}$.

This principle is quite useful in explaining a number of phenomena. For example, the change in angular velocity of a spinning ice skater can be explained by using the law of conservation of angular momentum. As the skater brings her arms and legs close to her body, her moment of inertia decreases and as a result her angular velocity increases.

EXAMPLE PROBLEM 4. A student stands on a rotating platform that has frictionless bearings. He has a 2.00 kg object in each hand, held 1.00 m from the axis of rotation of the system. The system is initially rotating at 10.0 rpm. Determine the a) initial angular velocity of the system in radians per second, b) angular velocity of the system in radians per second if the objects are brought to a distance of 0.200 m from the axis of rotation, and c) change in the rotational kinetic energy of the system as the objects are pulled closer to the center of rotation. d) What causes the increase in the rotational kinetic energy? Assume that the moment of inertia of the platform + student remains constant at 1.00 kg m^2.

Part a. Step 1. Determine the initial angular velocity in radians per second.	Solution: (Section 8-8) 1 rev = 2 π radians and 1 minute = 60.0 sec ω = (10.0 rev/min)(2 π rad/rev)(1 min/60.0 s) = 1.05 rad/s
Part b. Step 1. Determine the initial and final moment of inertia of the system.	$I = \Sigma mr^2$ includes the moment of inertia of each object as well as the moment of inertia of the student + platform. $I = m_1 r_1^2 + m_2 r_2^2 + I_{student + platform}$ I_i = (2.00 kg)(1.00 m)^2 + (2.00 kg)(1.00 m)^2 + 1.00 kg m^2 = 5.00 kg m^2 I_f = (2.00 kg)(0.200 m)^2 + (2.00 kg)(0.200 m)^2 + 1.00 kg m^2 = 1.16 kg m^2
Part b. Step 2. Determine the final angular velocity of the system.	Since no external torque acts on the system, apply the law of conservation of angular momentum, $I\omega$ = constant. $I_i\ \omega_i = I_f\ \omega_f$ (5.00 kg m^2)(1.05 rad/s) = (1.16 kg m^2) ω_f

	ω_f = 4.53 rad/s
Part c. Step 1. Determine the change in KE of the system as the objects are brought closer to the axis of rotation.	$\Delta KE = KE_f - KE_i$ $\Delta KE = \frac{1}{2} I_f \omega_f^2 - \frac{1}{2} I_i \omega_i^2$ $\Delta KE = \frac{1}{2}(1.16 \text{ kg m}^2)(4.53 \text{ rad/s})^2 - \frac{1}{2}(5.00 \text{ kg m}^2)(1.05 \text{ rad/s})^2$ ΔKE = 11.9 J - 2.76 J = 9.14 J
Part d. Step 1. What causes the increase in the rotational kinetic energy?	As the student pulls the objects in toward the center of rotation, he is doing work on the system. The energy he expends in doing this work appears in the form of increased kinetic energy.

TEXTBOOK QUESTION 14. We claim that momentum and angular momentum are conserved. Yet most moving or rotating bodies eventually slow down and stop. Explain.

ANSWER: Momentum and angular momentum are conserved if no outside net force or net torque acts on the object or system of objects. A spinning ice skater eventually loses angular momentum and slows down due to air resistance and friction between the skate and the ice. A book sliding across a smooth horizontal table loses momentum due to friction between the bottom of the book and the table.

PROBLEM SOLVING SKILLS

For problems involving angular kinematics:

1. Complete a data table using information both given and implied in the problem.
2. Memorize the kinematics formulas for uniformly accelerated angular motion. Using the data from the completed data table, determine which formula or combination of formulas can be used to solve the problem.

For problems involving torque:

1. Draw a diagram locating the axis of rotation of the object.
2. Determine the magnitude of the force and the lever arm distance.
3. Solve for the magnitude of the torque. Use the right-hand rule to determine the direction of rotation and also the direction of the torque vector.

For problems involving torque and rotational dynamics:

1. Determine the moment of inertia of the object(s).
2. Use $\tau = \ell_\perp F$ and $\tau = I\alpha$ to solve for the object's angular acceleration.
3. If necessary, use the kinematics equations to solve for the angular velocity and linear velocity of the object(s).

If the problem involves a system of objects:

1. Draw a free body diagram and locate the forces acting on each object.
2. Use Newton's second law as well as $\tau = \ell_{\perp} F$ and $\tau = I \alpha$ to write an equation for each object.
3. Use the kinematics relationships as well as standard algebraic methods to solve for the rate of acceleration of the system.
4. If necessary, use the kinematics equations to solve the problem.

For problems involving the law of conservation of angular momentum:

1. If a net external torque acts on the system, the law of conservation of angular momentum does not apply and cannot be used.
2. Convert the angular velocity from revolutions per minute to radians per second.
3. Determine the initial and final moment of inertia of the system.
4. Use the law of conservation of angular momentum to solve the problem.
5. Calculate the change in rotational kinetic energy and explain why the kinetic energy changes while the angular momentum remains constant.

SOLUTIONS TO SELECTED TEXTBOOK PROBLEMS

TEXTBOOK PROBLEM 21. The tires of a car make 65 revolutions as the car reduces its speed from 95 km/h to 45 km/h. The tires have a diameter of 0.80 m. (a) What was the angular acceleration of the tires? (b) If the car continues to decelerate at this rate, how much more time is required for it to stop.

Part a. Step 1. Convert the car's speed to m/s.	Solution: (Section 8-2) v_o = (95 km/h)(1000 m/km)(1 h/3600 s) = 26.4 m/s v = (45 km/h)(1000 m/km)(3600 s/h) = 12.5 m/s
Part a. Step 2. Complete a data table based on information given in the problem.	Since the angular acceleration and displacement are to be expressed in radians, it is necessary to determine the initial and final angular velocities in radians per second and the angular displacement in radians. $\omega_o = v_o/r$ = (26.4 m/s)/(0.40 m) = 66.0 rad/s $\omega = v/r$ = (12.5 m/s)/(0.40 m) = 31.2 rad/s $\theta - \theta_o$ = (65 rev)(2π rad/rev) = 408 rad α = ?
Part a. Step 3. Solve for the angular acceleration.	$2\alpha(\theta - \theta_o) = \omega^2 - \omega_o^2$ $2(\alpha)$(408 rad) = $(31.2\ \text{rad/s})^2 - (66.0\ \text{rad/s})^2$ $\alpha = -4.1\ \text{rad/s}^2$

Part b. Step 1.

Complete a data table using information applicable to part b and solve for the time.

The angular acceleration is now known and can be added to the data table.

ω_o = 31.2 rad/s ω = 0 rad/s α = - 4.14 rad/s^2 t = ?

$\omega = \alpha t + \omega_o$

0 rad/s = (- 4.14 rad/s^2) t + 31.2 rad/s

t = 7.6 s

TEXTBOOK PROBLEM 24. Calculate the net torque about the axle of the wheel shown in Fig. 8-39. Assume that a friction torque of 0.40 m· N opposes the motion.

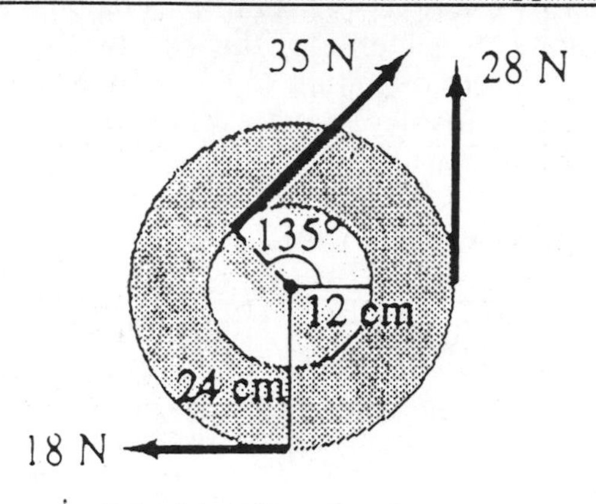

Part a. Step 1.

Determine the torque produced by each force, $\tau = \ell_\perp F$.
Note: r = 0.10 m and R= 0.20 m.

Solution: (Section 8-4)

$\tau_{18\ N}$ = (0.24 m)(18 N) sin 90° = 4.3 m·N (clockwise)

$\tau_{35\ N}$ = (0.12 m)(35 N) sin 90° = 4.2 m·N (clockwise)

$\tau_{28\ N}$ = (0.24 m)(28 N) sin 90° = 6.7 m·N (counterclockwise)

Part a. Step 2.

Determine the net torque acting on the wheel.

The sum of the clockwise torques is greater than the counterclockwise torque. Therefore, the frictional torque is in the counterclockwise direction. Let the clockwise direction represent the positive direction.

Net τ = 4.3 m·N + 4.2 m·N - 6.7 m·N - 0.40 m·N

Net τ = 1.4 m·N in the clockwise direction

TEXTBOOK PROBLEM 36. A teenager pushes tangentially on a small hand-driven merry-go-round and is able to accelerate it from rest to a frequency of 15 rpm in 10.0 s. Assume the merry-go-round is a uniform disk of radius 2.5 m and has a mass of 760 kg, and two children (each with a mass of 25 kg) sit opposite each other on the edge. Calculate the torque required to produce the acceleration, neglecting frictional torque. What force is required at the edge?

Part a. Step 1. Convert rpm to radians per second.	Solution: (Sections 8-5 and 8-6) $(15\ \text{rpm})(2\pi\ \text{rad/1rev})(1\ \text{min/60 s}) = 1.57\ \text{rad/s}$
Part a. Step 2. Determine the angular acceleration.	$\omega = \alpha\ t + \omega$ $1.57\ \text{rad/s} = \alpha\ (10.0\ \text{s}) + 0$ $\alpha = 0.157\ \text{rad/s}^2$
Part a. Step 3. Determine the moment of inertia of the disk + children.	$I_{total} = I_{disk} + I_{child} + I_{child}$ but $I_{disk} = \frac{1}{2}\ m_{disk}\ r^2$ and $I_{child} = m_{child}\ r^2$ $= \frac{1}{2}\ (760\ \text{kg})(2.5\ \text{m})^2 + 2\ (25\ \text{kg})(2.5\ \text{m})^2$ $I_{total} = 2380\ \text{kg m}^2 + 313\ \text{kg m}^2 = 2690\ \text{kg m}^2$
Part a. Step 4. Determine the magnitude of the torque.	$\tau = I\ \alpha$ $= (2690\ \text{kg m}^2)(0.157\ \text{rad/s}^2)$ $\tau = 420\ \text{m N}$
Part a. Step 5. Determine the magnitude of the force.	$\tau = r\ F \sin\theta$ $420\ \text{m N} = (2.5\ \text{m})\ F \sin 90^\circ$ $F = 170\ \text{N}$

TEXTBOOK PROBLEM 49. Two masses, $m_1 = 18.0$ kg and $m_2 = 26.5$ kg, are connected by a rope that hangs over a pulley (Fig. 8-47). The pulley is a uniform cylinder of radius 0.260 m and mass 7.50 kg. Initially m_1 is on the ground and m_2 rests 3.00 m above the ground. If the system is now released, use conservation of energy to determine the speed of m_2 just before it strikes the ground. Assume the pulley is frictionless.

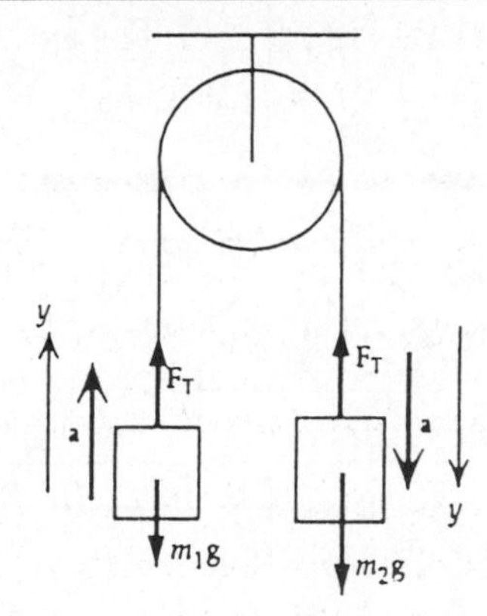

Part a. Step 1. Determine the moment of inertia of the pulley.	Solution: (Section 8-7) The pulley is a solid cylinder. $I_{pulley} = ½ \, m \, r^2 = ½ \, (7.50 \text{ kg})(0.260 \text{ m})^2 = 0.254 \text{ kg m}^2$
Part a. Step 2. Solve for the final velocity v_2' of m_2 using the law of conservation of energy.	The initial KE of each object = 0 J. The initial PE of m_1 = 0 J. The PE of the pulley does not change. The final PE of m_2 = 0 J. The final velocity of each object as well as a point on the rim of the pulley is the same. $(KE_1 + PE_1 + KE_2 + PE_2 + KE_{pulley} + PE_{pulley})_{initial} =$ $(KE_1 + PE_1 + KE_2 + PE_2 + KE_{pulley} + PE_{pulley})_{final}$ $0 \text{ J} + 0 \text{ J} + 0 \text{ J} + m_2 \, g \, y_2 + 0 \text{ J} + m_2 \, g \, y_{pulley} =$ $½ \, m_1 \, v_1'^2 + m_1 \, g \, y'_1 + ½ \, m_2 \, v_2'^2 + 0 \text{ J} + ½ \, I_{pulley} \, \omega^2 + 0 + m_2 \, g \, y'_{pulley}$ but, $m_2 \, g \, y_{pulley} = m_2 \, g \, y'_{pulley}$ and $v_1' = v_2'$ also, $½ \, I_{pulley} \, \omega'^2 = ½ \, (½ \, m_{pulley} \, r^2) \, \omega^2$ but $r^2 \, \omega^2 = v_2'^2$ therefore, $½ \, I_{pulley} \, \omega'^2 = ¼ \, m_{pulley} \, v_2'^2$ Substituting gives $m_2 \, g \, y_2 = ½ \, (m_1 + m_2) \, v_2'^2 + m_1 \, g \, y'_1 + ¼ \, m_{pulley} \, v_2'^2$ $(26.5 \text{ kg})(9.8 \text{ m/s}^2)(3.0 \text{ m}) =$ $½(18.0 \text{ kg} + 26.5 \text{ kg})v_2'^2 + (18.0 \text{ kg})(9.8 \text{ m/s}^2)(3.0 \text{ m}) + ¼(7.50 \text{ kg})v_2'^2$ $780 \text{ J} = (22.3 \text{ kg})v_2'^2 + 530 \text{ J} + (1.88 \text{ kg})v_2'^2$ $250 \text{ J} = (24.2 \text{ kg}) \, v_2'^2$ $v_2'^2 = 10.4 \text{ m}^2/\text{s}^2$ $v_2' = 3.22 \text{ m/s}$

TEXTBOOK PROBLEM 57. (a) What is the angular momentum of a figure skater spinning (with arms close to her person) at 3.5 rev/s, assuming her to be a cylinder of height of 1.5 m, radius of 15 cm, and a mass of 55 kg? (b) How much torque is required to slow her to a stop in 5.0 s, assuming she does *not* move her arms.

Part a. Step 1. Determine the angular velocity in radians per second.	Solution: (Section 8-8) 1 rev = 2 π radians $\omega = (3.5 \text{ rev/s})(2 \, \pi \text{ rad/rev}) = 22 \text{ rad/s}$

Part a. Step 2. Determine her moment of inertia.	Assume that her body approximates a solid cylinder 1.5 m high, with a radius of 15 cm (0.15 m), and a mass of 55 kg. $I_{skater} \approx I_{solid\ cylinder} = \frac{1}{2} m r^2 = \frac{1}{2} (55\ kg)(0.15\ m)^2 \approx 0.62\ kg\ m^2$
Part a. Step 3. Determine the skater's angular momentum.	$L = I\ \omega \approx (0.62\ kg\ m^2)(22\ rad/s)$ $L \approx 14\ kg\ m^2/s$
Part b. Step 1. Determine the angular deceleration.	$\omega = \omega_o + \alpha\ t$ $0\ rad/s = 22\ rad/s + \alpha\ (5.0\ s)$ $\alpha = -\ 4.4\ rad/s^2$
Part b. Step 2. Determine the magnitude of the torque.	$\tau = I\ \alpha$ $\tau \approx (0.62\ kg\ m^2)(-\ 4.4\ rad/s^2) \approx -\ 2.7\ N\ m$ Alternate: The torque equals the change in the angular momentum. $\tau = \Delta L/\Delta t \approx (0\ km\ m^2 - 14\ kg\ m^2)/(5.0\ s) \approx -\ 2.7\ N\ m$

TEXTBOOK PROBLEM 84. A marble of mass m and radius r rolls along the looped rough track of Fig. 8-58. What is the minimum value of *h* if the marble is to reach the highest point of the loop without leaving the track? Assume that $r \ll R$, and ignore frictional losses.

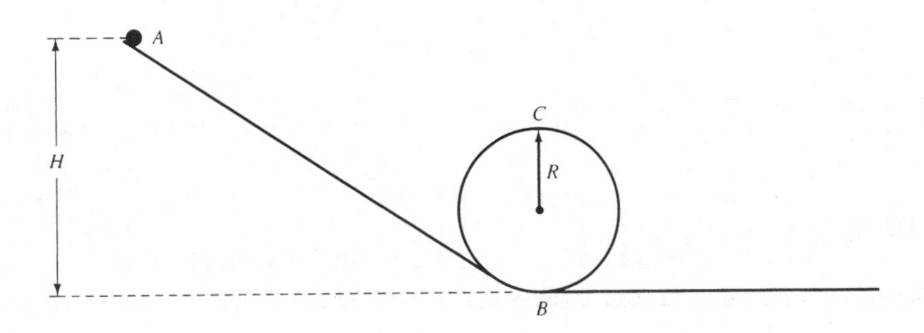

Part a. Step 1. Determine the ball's rate of acceleration at the top of the loop.	Solution: (Section 8-7) At the top of the loop the marble is just "touching" the track but the track exerts no force on the marble. The net force on the marble is the gravitational force. The gravitational force produces a centripetal acceleration and $a_R = g$.

Part a. Step 2. Derive a formula for the object's speed at the top of the loop.	$a_c = g = v^2/R$ solving for v gives $v^2 = g\ R$ $v = (R\ g)^{1/2}$
Part a. Step 3. Determine the total energy at the release point.	At release point h, the ball is at rest, therefore, KE = 0 while the PE = m g h. $\text{energy}_{\text{total at release point}} = m\ g\ h$
Part a. Step 4. Determine the total energy at the top of the loop.	At top of loop, the marble will be 2R above the bottom of the track. Therefore, $PE_{\text{top of loop}} = m\ g\ (2R)$ The kinetic energy at the top of the loop is the sum of the translational kinetic energy and the rotational kinetic energy. $KE_{\text{top of loop}} = KE_{\text{translation}} + KE_{\text{rotation}}$ $KE_{\text{top of loop}} = \frac{1}{2}\ m\ v^2 + \frac{1}{2}\ I\ \omega^2$ and $I_{\text{marble}} = (2/5)\ mr^2$ $= \frac{1}{2}\ m\ v^2 + \frac{1}{2}\ (2/5)\ mr^2\ \omega^2$ $= \frac{1}{2}\ m\ v^2 + (1/5\ mr^2)\ \omega^2$ but $r^2\ \omega^2 = v^2$ $KE_{\text{top of loop}} = \frac{1}{2}\ m\ v^2 + (1/5)\ mv^2 = (7/10)\ m\ v^2$ $\text{energy}_{\text{total at top of loop}} = (PE + KE)_{\text{top of loop}}$ $\text{energy}_{\text{total at top of loop}} = m\ g\ (2R) + (7/10)\ m\ v^2$
Part a. Step 4. Use the law of conservation of energy to solve for the minimum value of h.	total energy at the release point = total energy at top of track $m\ g\ h = m\ g\ (2R) + (7/10)\ m\ v^2$ The mass appears in each term. It cancels algebraically from each side of the equation. $g\ h = 2\ g\ R + (7/10)\ v^2$ But as shown in Step 2, at the top of loop $v^2 = R\ g$ therefore, $g\ h = 2\ g\ R + (7/10)\ g\ R$ Note: g cancels $h = 2\ R + (7/10)\ R$ $h = 2.7\ R$

CHAPTER 9

STATIC EQUILIBRIUM; ELASTICITY AND FRACTURE

OBJECTIVES

After studying the material of this chapter, the student should be able to:

- distinguish between static and dynamic equilibrium and state the two conditions for equilibrium.
- solve equilibrium problems using the two conditions for equilibrium.
- calculate the IMA, AMA, and efficiency of a simple machine.
- state whether an object is in stable, unstable, or neutral equilibrium.
- use Hooke's law to the solve problems.
- distinguish between stress and strain and between tensile stress, compressive stress, and shear stress.
- write the equations for the relationship between stress and strain for the three types of deformation of an elastic solid. Use these equations to solve problems.

KEY TERMS AND PHRASES

equilibrium occurs when an object is not accelerated. In order for an object to be in equilibrium the following two conditions for equilibrium must be satisfied.

first condition of equilibrium is satisfied when the vector sum of the forces acting on an object equals zero, $\Sigma \mathbf{F} = 0$.

second condition of equilibrium is satisfied when the sum of all torques acting on an object about any axis perpendicular to the plane of the forces equals zero, i.e., $\Sigma \tau = 0$.

static equilibrium occurs when an object at equilibrium is also at rest.

dynamic equilibrium occurs when an object at equilibrium is moving in a straight line at a constant speed.

concurrent forces are aligned such that the line of action of each force acting on the object passes through a common point. If the forces are concurrent, only the first condition of equilibrium ($\Sigma \mathbf{F} = 0$) is needed to solve the problem.

nonconcurrent forces are aligned such that the line of action of each force does not pass through a common point. The result is a net torque which tends to cause the object to rotate. If the forces are nonconcurrent, and yet the object is in equilibrium, then both conditions of equilibrium are required to solve the problem, i.e., $\Sigma\vec{\mathbf{F}} = 0$ and $\Sigma\tau = 0$.

machines are devices used to change the magnitude and/or direction of a force by changing the distance over which the force is exerted. In the process some practical advantage is achieved. Examples of simple machines include pulleys, levers, and incline planes.

ideal mechanical advantage (IMA) is the advantage if the work done by friction equals zero.

actual mechanical advantage (AMA) is the advantage when the work done by friction is included.

efficiency of a machine is defined as the amount of useful work output from the machine divided by the amount of work input. The efficiency also equals the ratio of the actual mechanical advantage to the ideal mechanical advantage.

stable equilibrium occurs when a small displacement from its undisturbed position results in a torque that tends to restore the object to its original position. An object remains in stable equilibrium as long as the object's center of gravity remains within its base of support.

unstable equilibrium occurs when any displacement from the object's undisturbed position results in a torque that tends to cause the object to move farther away from its original position. An example is a pencil balanced vertically on its point.

neutral equilibrium occurs if the object remains in stable equilibrium independent of the object's orientation. An example is a spherical ball placed on a flat surface. No matter how the ball is displaced the cg remains over the support point.

Hooke's law states that the deformation of an object is proportional to the magnitude of the applied force and a proportionality constant called the force constant.

stress is the ratio of the force causing an object's distortion to the area on which the force acts.

strain is the ratio of the object's change in length to the object's original length.

fracture occurs when the stress on an object exceeds the elastic limit of the object. The force exerted per unit area at the fracture point is called the ultimate strength of the material.

SUMMARY OF MATHEMATICAL FORMULAS

first condition of equilibrium	$\Sigma\vec{\mathbf{F}} = 0$	The vector sum of the forces acting on an object equals zero, $\Sigma\vec{\mathbf{F}} = 0$.
second condition of equilibrium	$\Sigma\tau = 0$	The vector sum of all torques acting about any axis perpendicular to the plane of the forces equals zero, i.e., $\Sigma\tau = 0$.

ideal mechanical advantage (IMA)	IMA = $(d_i)/(d_o)$	The "ideal" mechanical advantage (IMA) is the advantage if the work done by friction equals zero. IMA is the ratio of the distance (d_i) moved by the input force to the distance (d_o) moved by the output force.
actual mechanical advantage (AMA)	AMA = F_o/F_i	The actual mechanical advantage (AMA) is the advantage when the work done by friction is included. AMA is the ratio of the output force (F_o) to the input force (F_i).
efficiency	$e = (W_o)/(W_i)$ $e = (F_o\ d_o)/(F_i\ d_i)$ $e = [(F_o)/(F_i)]/[(d_i)/(d_o)]$ e = AMA/IMA	The efficiency (e) of a machine is defined as the amount of work output from the machine divided by the amount of work input. The efficiency can also be written as the ratio of the actual mechanical advantage (AMA) to the ideal mechanical advantage (IMA).
Hooke's law	$\Delta F = k\ \Delta L$	The deformation (ΔL) of an elastic object is related to the magnitude of the applied force (ΔF) and the force constant (k).
tensile and compressive stress	$\Delta F/A = E\ (\Delta L/L_o)$	Tensile (or compressive) stress ($\Delta F/A$) is the force per unit area on which the force acts. Strain is the ratio of the change in length to the original length; strain = $\Delta L/L_o$. The stress is related to the strain by elastic modulus (E).
shear stress	$\Delta F/A = G\ (\Delta L/L_o)$	Shear stress ($\Delta F/A$) occurs when equal but opposite forces are applied tangentially across the opposite faces of an object. Shear stress is related to the strain ($\Delta L/L_o$) by the shear modulus (G).
volume stress	$\Delta P = -B\ (\Delta V/V_o)$	A compressive stress (ΔP) that acts over the entire surface of an object will cause a decrease in the object's volume (ΔV). V_o is the original volume and B is a proportionality constant called the bulk modulus. The negative sign indicates that the volume of the object decreases with increasing pressure.

CONCEPT SUMMARY

A body is in equilibrium if it is at rest (**static equilibrium**) or moving in a straight line at a constant speed (**dynamic equilibrium**).

First Condition of Equilibrium

In order for an object to be in equilibrium, the vector sum of the forces acting on it must be zero, $\Sigma\vec{F} = 0$. To solve problems involving equilibrium, the force vectors are resolved into x, y, and z components. The equation $\Sigma\vec{F} = 0$ is replaced by $\Sigma\vec{F}_x = 0$, $\Sigma\vec{F}_y = 0$, and $\Sigma\vec{F}_z = 0$.

If the line of action of each force acting on the object passes through a common point, the forces are termed **concurrent** and $\Sigma\vec{F} = 0$ is the only condition needed to solve the problem. Below is an example of forces acting on an object which are concurrent.

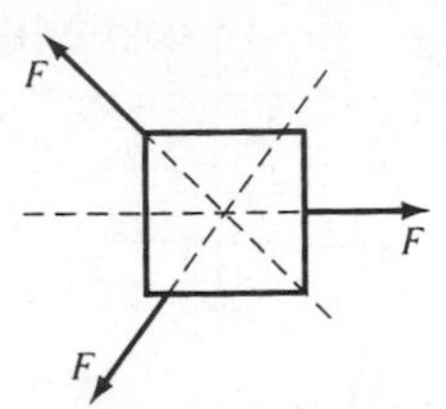

TEXTBOOK QUESTION 2. A bungee jumper momentarily comes to rest at the bottom of the dive before he springs back upward. At that moment is the bungee jumper in equilibrium? Explain.

ANSWER: When the bungee jumper starts his dive, his weight is greater than the upward force exerted by the cord and he accelerates downward. As he travels downward, the upward force exerted by the cord gradually increases until it exceeds the diver's weight and the diver decelerates to a momentary halt. At that moment he is not in equilibrium because the upward force exerted by the cord is greater than his weight. Using Newton's second law, the net force, i.e., $F_{cord} - mg = ma$ and as a result the diver accelerates upward.

EXAMPLE PROBLEM 1. A 15.0 kg child holds on to a rope attached to a tree. The child's brother applies a horizontal force and pulls the child back a certain distance before releasing the child. Calculate the magnitude of the force exerted by the brother when the rope makes a 37.0° with the vertical.

Part a. Step 1. Draw an accurate diagram locating the forces acting on the rope.	Solution: (Section 9-2) 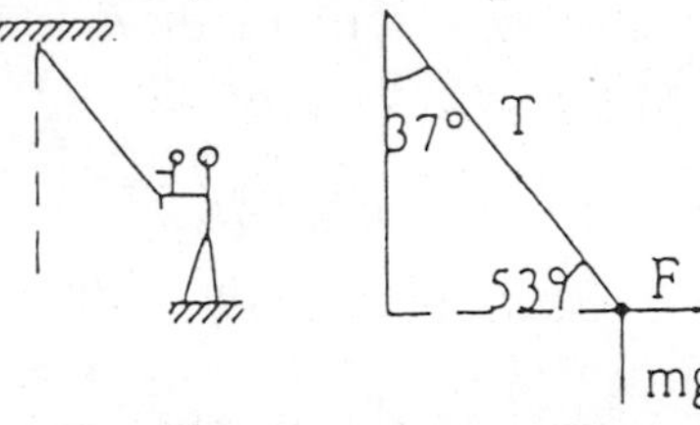
Part a. Step 2. Apply the first condition of equilibrium to the vertical components of force and solve for T.	$\Sigma F_y = 0 \qquad T_y - m\,g = 0$ $T \sin 53.0° - (15.0\text{ kg})(9.8\text{ m/s}^2) = 0$ $(0.799)\,T - 147\text{ N} = 0$ $T = (147\text{ N})/(0.799) = 184\text{ N}$

Part a. Step 3.	$\Sigma F_x = 0 \qquad F - T_x = 0$
Apply the first condition of equilibrium to the horizontal components of force and solve for F.	$F - T \cos 53.0° = 0$ $F - (184\ N)(0.602) = 0$ $F = 111\ N$

Second Condition of Equilibrium

If the forces are **nonconcurrent**, that is, the line of action of each force does not pass through a common point, then a net torque will cause a change in the state of rotation of the object. As can be seen in the diagram below, the object would tend to rotate in a clockwise direction.

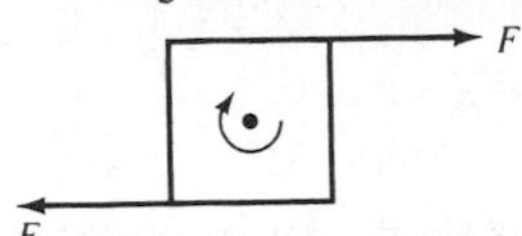

The **second condition of equilibrium** states that the vector sum of all torques acting about any axis perpendicular to the plane of the forces must be zero, i.e., $\Sigma \tau = 0$. A torque that causes a counterclockwise rotation is defined as positive while a clockwise torque is defined as negative. Thus, for equilibrium, the sum of clockwise and counterclockwise torques must add to zero. In other words, the sum of the clockwise torques about any axis equals the sum of the counterclockwise torques about the axis.

Since no rotation occurs, any point may be selected to be the location of the reference axis. In order to simplify the solution to most problems, the axis of rotation is usually taken to be a point through which the line of action of an unknown force passes. In this situation, the lever arm distance from the rotation point to the line of action of the unknown force is zero. Therefore, the torque produced by the unknown force is zero and at least one unknown has been eliminated from the torque equation. For nonconcurrent force problems, both conditions of equilibrium must usually be applied to solve the problem.

EXAMPLE PROBLEM 2. During a classroom demonstration, the instructor asks Andy and Bob to lift a table on which a third student, Chuck, is sitting. The table is 2.0 m long and weighs 150 N. The table is uniform and it is held so that it remains horizontal even though Chuck is sitting on it. Chuck weighs 700 N and is sitting 0.60 m from Andy. Determine the force exerted by Andy and Bob.

Part a. Step 1.	Solution: (Sections 9-2 and 9-3)
Draw an accurate free body diagram locating the forces acting on the table.	Let F_A be the force that Andy exerts and F_B be the force exerted by Bob. In order to lift the table both students exert an upward force as shown in the diagram.

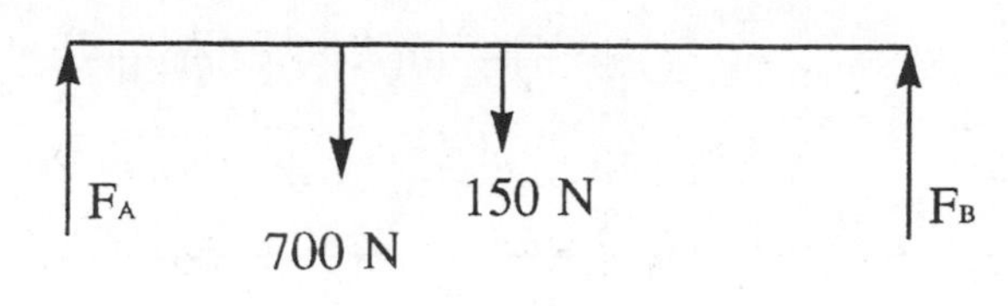

Part a. Step 2. Apply the first condition of equilibrium, $\Sigma F = 0$.	$\Sigma F_x = 0$ $\Sigma F_y = 0$	No horizontal forces are present in this problem. $F_A + F_B - 700\ N - 150\ N = 0$ $F_A + F_B = 850\ N$ (equation 1)

Part a. Step 3.

Apply the second condition of equilibrium, $\Delta\tau = 0$. Choose the center of the table as the axis of rotation.

In order to determine whether the torque produced by a force about the rotation point is clockwise (CW) or counterclockwise (CCW), let us use the tip of pencil method. Place the point of a pencil at the axis of rotation and hold it fixed. The rest of the pencil is parallel to the table on which the forces act. Applying the force to the pencil at the point in the diagram where the force is the located will cause the pencil to rotate either CW or CCW.

For example, place the point of the pencil at the center of the line representing the table. Pushing on the pencil at the point where Andy's force is located and in the direction of Andy's force causes the pencil to rotate CW. Thus, Andy's force causes a clockwise torque about the center of the board. Now arrange the pencil so that Bob's force acts. Note that this force produces a counterclockwise torque. Repeating this process we find that Chuck's weight produces a CCW torque. The weight of the board creates no torque. The weight is at the rotation point. The lever arm distance is zero; therefore, the torque is zero.

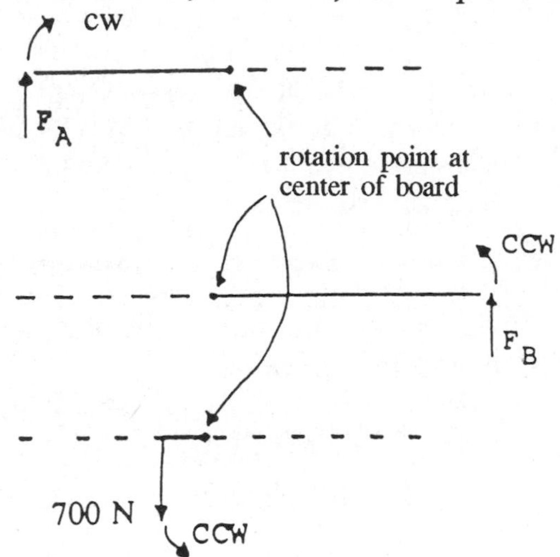

Since no rotation occurs,

$$\Sigma\tau_{CCW} - \Sigma\tau_{CW} = 0$$

$$F_B\ (1.0\ m) + (700\ N)(0.40\ m) - F_A\ (1.0\ m) = 0$$

Simplifying: $F_B - F_A = -280\ N$ (equation 2)

Part a. Step 4.

There are now two equations and the same two unknowns. Solve for each unknown.

Adding the two equations:

$F_A + F_B + F_B - F_A = 850\ N - 280\ N$

$2\ F_B = 570\ N$

$F_B = 285\ N$

Substituting $F_B = 285\ N$ into equation 1,

$F_A + 285\ N = 850\ N$, then

$F_A = 565\ N$

Part a. Step 5.

The choice of an axis of rotation is arbitrary. Select the left end and determine the force exerted by Andy and Bob.

It is convenient to select a point for the axis of rotation where an unknown force acts, e.g., the left end. Drawing the free body diagram and using the point of pencil method, we determine that the 700 N and 150 N forces create a CW torque, F_B produces a CCW torque, while F_A produces no torque since its lever arm distance equals zero.

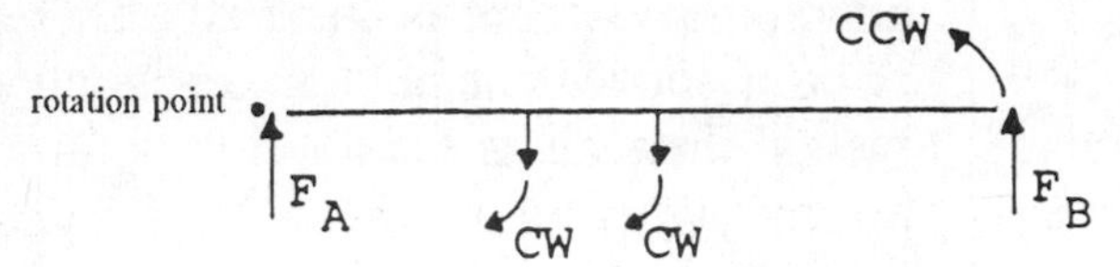

Applying the second condition of equilibrium:

$\Sigma\tau_{CCW} - \Sigma\tau_{CW} = 0$

$F_B(2.0\ m) - (700\ N)(0.60\ m) - (150\ N)(1.0\ m) = 0$

$F_B = 285$

To determine F_A, substitute the value of F_B into equation. 1.

$F_A + 285\ N = 850\ N$

$F_A = 565\ N$

EXAMPLE PROBLEM 3. A uniform pole 6.0 meters long weighs 800 N and is attached at one end to a vertical wall. A load of 400 N hangs from the outer end of the pole. A horizontal guy wire attached to the outer end of the pole holds the pole at a 37.0° angle with the horizontal. Determine the a) tension in the wire, and b) horizontal and vertical components of the force exerted by the wall on the pole.

Part a. Step 1.

Draw an accurate diagram locating each of the forces acting on the pole.

Solution: (Section 9-3)

The pole is uniform; therefore, the weight is concentrated at the center.

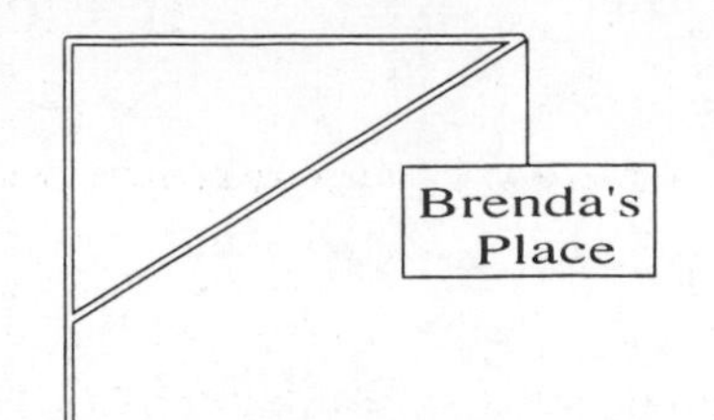

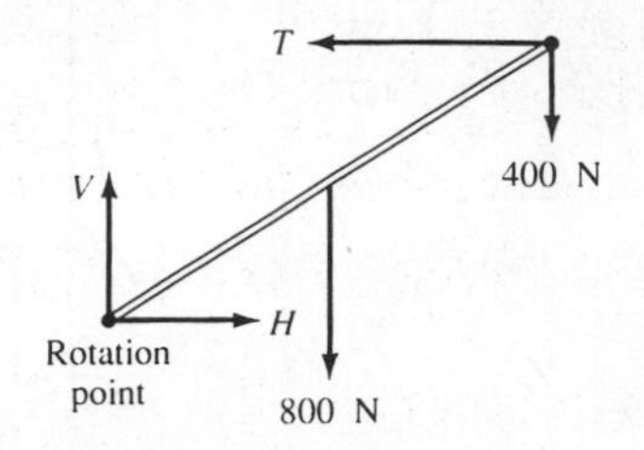

Part a. Step 2.

Apply the first condition of equilibrium.

$\Sigma F_x = 0$ $\quad H - T = 0$

$\Sigma F_y = 0$ $\quad V - 800\ N - 400\ N = 0$

$V = 1200\ N$

Part a. Step 3.

Apply the second condition of equilibrium, $\Sigma\tau = 0$, and solve for the tension in the wire. Note: choose the point where the pole is attached to the wall as the rotation point.

The convenient point for the axis of rotation is a point where an unknown force is located. Therefore, either end of the pole is be a convenient point. Let us choose the end of the pole that meets the wall as the rotation point. At this point the wall exerts a force which is represented by the components, H and V.

Use the point of pencil method to determine the direction of each torque. From the diagram it can be seen that T produces a CCW torque while the 800 N weight of the pole and the 400 N load each produce a CW torque. Since V and H act at the rotation point, the lever arm distance is zero and they produce no torque.

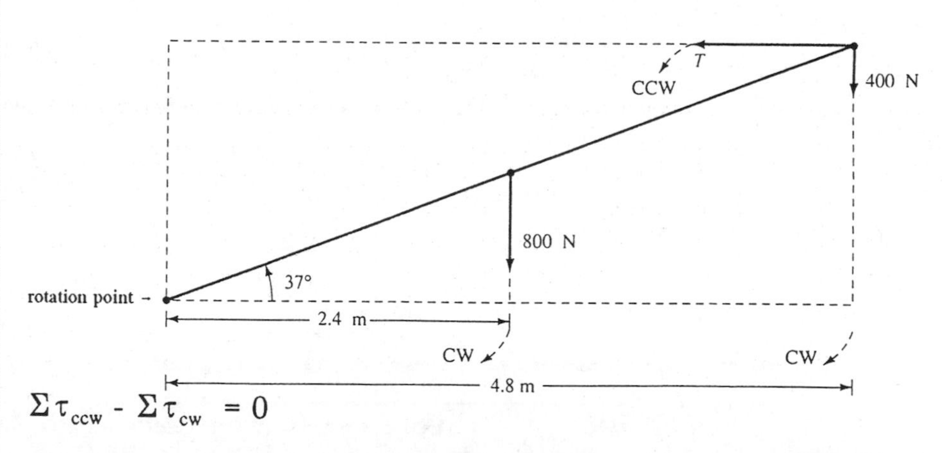

$$\Sigma\tau_{ccw} - \Sigma\tau_{cw} = 0$$

$$T(6.00m)(\sin 37.0°)(\sin 90°) - (800\ N)(3.00\ m)(\cos 37.0°)(\sin 90°) - (400\ N)(6.00\ m)(\cos 37.0°)(\sin 90°) = 0$$

$$(3.60\ m)\ T - 1920\ m\ N - 1920\ m\ N = 0$$

$$T = 1070\ N$$

Part b. Step 1. The value of V was obtained in Part a. Step 2. T is now known; solve for H.	From Part a. Step 2. $H - T = 0$ $H - 1070 \text{ N} = 0$ $H = 1070 \text{ N}$

Simple Machines: Levers and Pulleys

Machines are devices used to change the magnitude and/or direction of a force by changing the distance over which the force is exerted. In the process some practical advantage is achieved. Examples of simple machines include pulleys, levers, and incline planes.

The principle of work applies to machines. The work input is equal to the sum of the useful work output plus the work done by friction. When discussing machines it is convenient to refer to the mechanical advantage. The **"ideal" mechanical advantage** (IMA) is the advantage if the work done by friction equals zero. The IMA can be determined by using the following formula:

$IMA = (d_i)/(d_o)$ where d_i is the distance moved by the input force and d_o is the distance moved by the output force.

The **actual mechanical advantage** (AMA) is the advantage when the work done by friction is included. The formula for AMA is

$AMA = (F_o)/(F_i)$ where F_o is the output force acting on the object and F_i is the input force exerted by the object causing the motion.

The **efficiency** (e) of a machine is defined as the amount of work output from the machine divided by the amount of work input.

efficiency = e = work output/work input

$\text{efficiency} = e = (W_o)/(W_i) = (F_o\, d_o)/(F_i\, d_i) = [(F_o)/(F_i)]/[(d_i)/(d_o)] = AMA/IMA$

EXAMPLE PROBLEM 4. The machine shown in the diagram consists of two wheels that move together and are mounted on the same axle. The radii of the wheels are 0.10 m and 0.20 m, respectively. A 300 N force must be applied at F in order to raise 400 N at W. Determine the a) IMA, b) AMA, and c) efficiency of the machine.

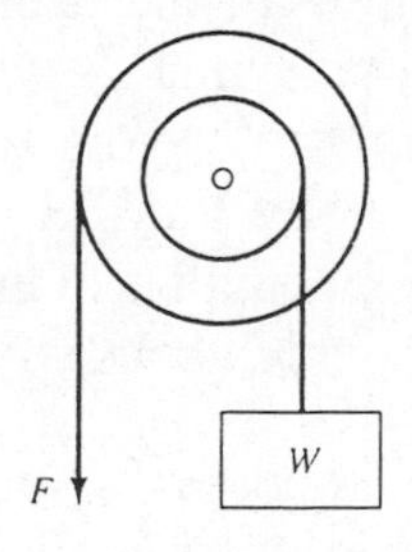

Part a. Step 1. Determine the mechanical advantage.	Solution: (Section 9-4) In one complete turn, a length of string equal to the circumference ideal of each wheel is wound or unwound. Apply the formula for IMA. IMA = $(d_i)/(d_o)$ where $d_i = 2\ \pi\ R$ and $d_o = 2\ \pi\ r$ $= (2\ \pi\ R)/(2\ \pi\ r) = (0.20\ m)/(0.10\ m)$ IMA = 2.0
Part b. Step 1. Determine the actual mechanical advantage.	$F_o = W = 400\ N$ and $F_i = F = 300\ N$ AMA = $(F_o)/(F_i)$ = (400 N)/(300 N) AMA = 1.3
Part c. Step 1. Determine the efficiency (e).	e = AMA/IMA X 100% e = [(1.3)/(2.0)] x 100% = 65%

Stable, Unstable, and Neutral Equilibrium

An object whose center of gravity is below its support point is said to be in **stable equilibrium**. Any displacement from its undisturbed position results in a torque that tends to restore it to its original position. An example of this is a ball on the end of a string. When displaced from its undisturbed position, a component of its weight ($F_{\parallel}$) tends to return the ball to its original orientation.

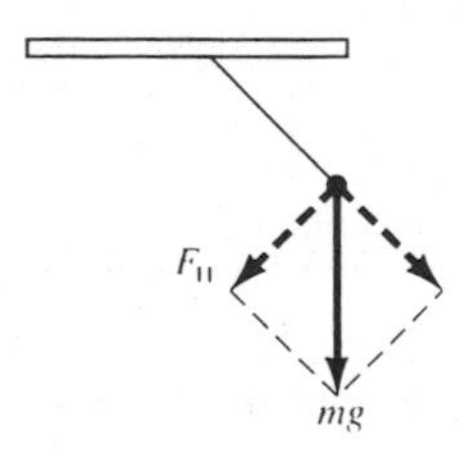

If the center of gravity is above the support point, the question of stable equilibrium depends on the location of the position of the center of gravity (CG) relative to the support point. As stated in the textbook, "In general, a body whose cg is above its base of support will be stable if a vertical line projected downward from its cg falls within the base of support." To clarify this idea try the following experiment. Stand with the heels of your shoes against a wall. If you slowly bend over to touch your toes with your fingers without bending your knees, you will find your toes gradually supporting more and more of your weight. Past a certain point you will fall forward. This is because past a certain point the CG of the upper portion of your body falls outside your shoes, which form your base of support. Once your center of gravity is outside your shoes, an unbalanced torque is produced which causes you to topple over.

If you move away from the wall and bend over, you will not fall forward. This is because your hips move backward as the upper portion of your body moves forward. Thus your CG remains within your base of support and you will be able to touch your toes.

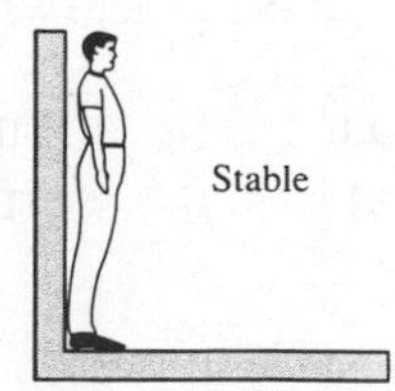

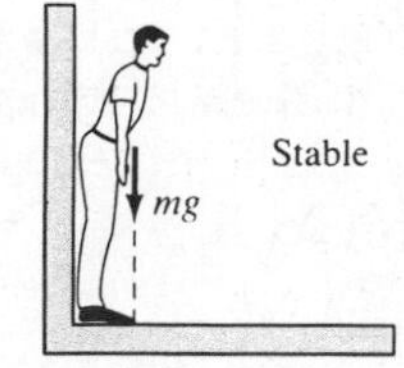

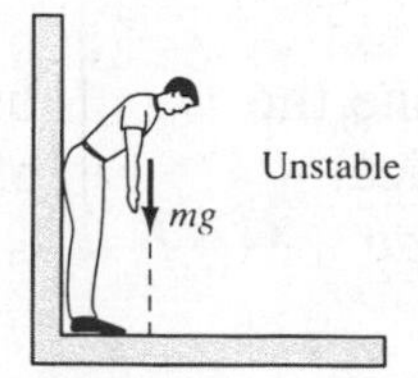

An example of an object in neutral equilibrium is a spherical ball placed on a flat surface. No matter how the ball is displaced the CG remains over the support point.

TEXTBOOK QUESTION 12. Why do you tend to lean backward when carrying a heavy load in your arms?

ANSWER: A heavy load in your arms tends to move the center of gravity of your body + load outward from the center of your body. If a line from the center of gravity to the floor passes within the base of your feet, you will be in equilibrium. If the line passes beyond your toes, you will tend to tip forward and fall on your face. By leaning backward you are shifting the center of mass of your body + load so that it remains within the base of your feet.

TEXTBOOK QUESTION 13. Place yourself facing the edge of an open door. Position your feet astride the door with your nose and abdomen touching the door's edge. Try to rise on your tiptoes. Why can't this be done?

ANSWER: In order to rise onto your tiptoes, the center of mass of your body must be shifted forward until it is over your toes. Since your nose and abdomen are already touching the door's edge, it is impossible to shift your weight forward. You feel that your feet are nailed to the floor.

Note: if you try to do this you will find that you may be able to momentarily shift your weight but you immediately tend to fall back onto your heels.

Elasticity: Stress, Strain, and Fracture

Robert Hooke (1635-1703) first stated the relationship that connects the deformation of a body to the magnitude of the applied force.

$\Delta F = k\,\Delta L$ where ΔF is the applied force, ΔL is the deformation of the object, and k is the force constant in N/m.

Hooke's law is found to hold for most materials, e.g., a wire, rubber band, or a spring, up to a point which is called the elastic limit. Beyond this limit the object will not return to its original shape when the force is removed. An object deformed beyond the elastic limit will eventually reach its breaking point and fracture.

Tension, Compression, and Shear

Hooke's law can be stated in terms of **tensile stress** and **strain** and a proportionality constant called the **elastic modulus** (E), Young's modulus, or the tensile modulus.

Stress is defined as the force per unit area on which the force acts, stress = $\Delta F/A$. Strain is defined as the ratio of the change in length to the original length, strain = $\Delta L/L_o$ and elastic modulus = stress/strain.

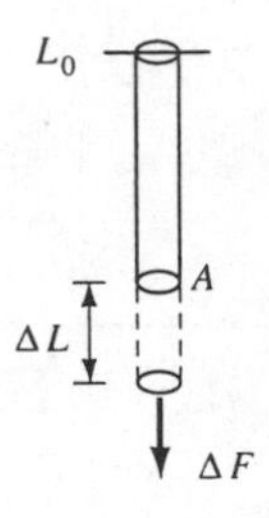

$E = (\Delta F/A)/(\Delta L/L_o)$ and $\Delta F/A = E\ (\Delta L/L_o)$

Compressive Stress

Compressive stress is the compression of a body by an outside force that acts inwardly on the object. The equation for tensile stress also applies to compressive stress, and the values of E are usually the same for both tensile stress and compressive stress.

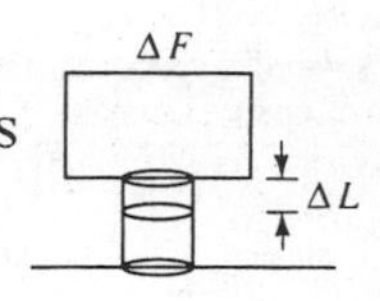

Shear Stress

Shear stress occurs when equal but opposite forces are applied tangentially across the opposite faces of an object.

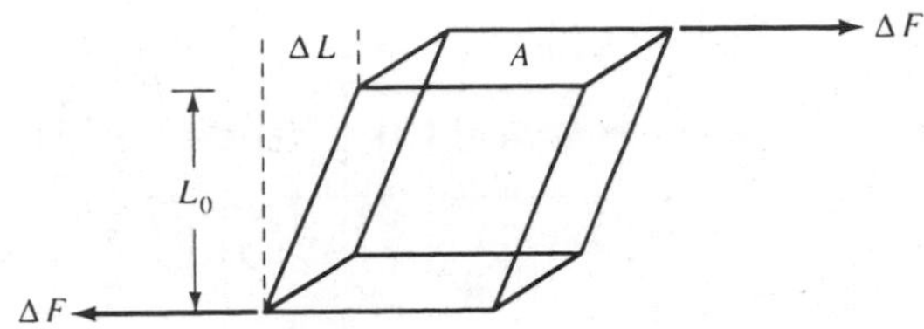

The shear modulus (G) equals the ratio of the shear stress to the shear strain and is given by the formula:

$G = (\Delta F/A)/(\Delta L/L_o)$ and $\Delta F/A = G\ (\Delta L/L_o)$

Table 9-1 in the textbook lists the values of E and G for a number of materials.

Fracture

If the stress on an object is too great, the object fractures or breaks. The force exerted per unit area at the fracture point is called the ultimate strength of the material.

EXAMPLE PROBLEM 5. A metal wire 0.50 m long and 1.00×10^{-2} meters in diameter stretches 3.2×10^{-4} m when a load of 10.0 kg is attached at the end. Determine the a) stress, b) strain, and c) elastic modulus for the material of the wire.

Part a. Step 1.	Solution: (Section 9-6)
Determine the tensile stress.	stress = $\Delta F/A$ and $A = \pi r^2$, where r = dia/2
	$r = (1.00 \times 10^{-2}\ m)/2 = 0.50 \times 10^{-2}\ m$
	stress = $[(10.0\ kg)(9.8\ m/s^2)]/[\pi(0.50 \times 10^{-2}\ m)^2]$
	stress = $1.25 \times 10^6\ N/m^2$

Part b. Step 1. Determine the strain.	strain $= \Delta L/L_o$ $= (3.20 \times 10^{-4}\ m)/(0.50\ m)$ strain $= 6.40 \times 10^{-4}$
Part c. Step 1. Determine the elastic modulus.	E = stress/strain $= (1.25 \times 10^{6}\ N/m^2)/(6.40 \times 10^{-4})$ $E = 1.95 \times 10^{9}\ N/m^2$

Bulk Modulus

A compressive stress that acts over the entire surface of an object will cause a decrease in the object's volume. If the force per unit area is uniform, the relationship that connects the volume stress and the change in volume is given by

$B = - \Delta P/(\Delta V/V_o)$ and $\Delta P = - B\ (\Delta V/V_o)$

where ΔP is the volume stress, which for a liquid is the change in pressure. ΔV is the change in volume. The negative sign appears because the volume of the object decreases with increasing pressure. V_o is the original volume and B is a proportionality constant called the bulk modulus.

PROBLEM SOLVING SKILLS

For problems where the forces are concurrent and the object is in static equilibrium:

1. Draw an accurate diagram locating the forces acting on the object or system of objects.
2. Draw a free body diagram locating the forces acting on the object(s) in question.
3. Resolve each force vector into x and y components.
4. Apply the first condition of equilibrium ($\Sigma F_x = 0$, $\Sigma F_y = 0$) and solve the problem.

For problems where the forces are nonconcurrent and the object or system of objects is in static equilibrium:

1. Repeat steps 1, 2, and 3 listed above.
2. Write an equation(s) using the first condition of equilibrium.
3. Select a convenient point for the axis of rotation. A convenient point is located at a position where an unknown force acts.
4. Use the tip of pencil method, refer to example problem 2, to determine the direction of the rotation (CW or CCW) produced by the force about the axis of rotation.
5. Write an equation using the second condition of equilibrium.
6. Use the equations written using the two conditions of equilibrium to solve the problem.

For problems involving mechanical advantage and efficiency:

1. Draw a diagram locating the forces acting on the object.
2. Determine the input force, output force, input distance, and output distance.
3. Use the appropriate formula to determine the IMA, AMA, and efficiency.

For problems involving stress and strain:

1. Draw an accurate diagram locating the forces acting on the object.
2. Determine whether the problem involves tensile stress, compressive stress, or shear stress.
3. Apply the appropriate formula and solve the problem.

SOLUTIONS TO SELECTED TEXTBOOK PROBLEMS

TEXTBOOK PROBLEM 6. Calculate the forces F_A and F_B that the supports exert on the diving board of Fig. 9-42 when a 58 kg person stands at its tip. (a) Ignore the weight of the board. (b) Take into account the board's mass of 35 kg. Assume that the board's CG is at its center.

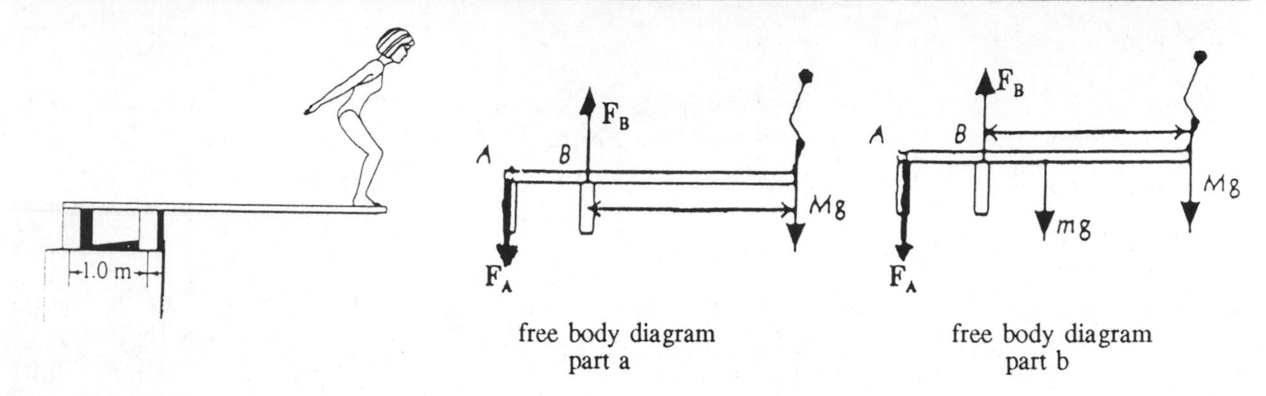

free body diagram part a

free body diagram part b

Part a. Step 1.

Draw an accurate free body diagram locating the forces acting on the board. Determine the direction of the force exerted by each support.

Solution: (Sections 9-1 and 9-2)

Let F_A represent the force that the support at the left end exerts and F_B the force exerted by the support 1.0 m from the left end. In order to determine the direction of the force exerted by each support, place a ruler on a table with the right edge of the ruler extended over the table. The weight of the student can be represented by pushing down on the outer edge of the ruler which extends over the table. As this is done you will note that the left end of the ruler must be held down (F_A) while the edge of the table exerts an upward force (F_B).

Part a. Step 2.

Apply the first condition of equilibrium, $\Sigma F = 0$.

$\Sigma F_x = 0$ No horizontal forces are present in this problem.

$\Sigma F_y = 0 \quad F_B - F_A - (58\text{ kg})(9.80\text{ m/s}^2) = 0$

$F_B - F_A = 5.7 \times 10^2\text{ N}$

Part a. Step 3.

Choose the left end the diving board as the rotation point.

Use the tip of pencil method to determine the direction (CW or CCW) of the torque produced

According to theory, since no rotation occurs, any point may be selected as the location of the reference point. Therefore, let the left end of the diving board be chosen as the rotation point. Using point of pencil method, place the point of the pencil at the left end of the line representing the board. Pushing on the pencil at the point where F_B is located causes the pencil to rotate CCW. Thus F_B causes a counterclockwise torque about the about the left end of the board. Now repeat for the student's weight. Note that the weight produces a clockwise torque (CW) about the left end of the board.

Part a. Step 4. Use the second condition of equilibrium to solve for the force exerted by support F_B.	$\Sigma\tau_{CCW} - \Sigma\tau_{CW} = 0$ F_B (1.0 m) sin 90° - (58 kg)(9.80 m/s^2)(4.0 m) sin 90° = 0 F_B (1.0 m) - 2270 N m = 0 $F_B = 2.3 \times 10^3$ N (upward)
Part a. Step 5. Apply the first condition of equilibrium, $\Sigma F = 0$, to solve for F_A.	From Part a. Step 2. $F_B - F_A = 570$ N but $F_B = 2270$ N 2270 N - F_A = 570 N $F_A = 1.7 \times 10^3$ N (downward)
Part b. Step 1. Apply the first condition of equilibrium, $\Sigma F = 0$.	$\Sigma F_x = 0$ No horizontal forces are present in this problem. $\Sigma F_y = 0$ $F_B - F_A$ - (58 kg)(9.80 m/s^2) - (35 kg)(9.80 m/s^2) = 0 $F_B - F_A = 9.1 \times 10^2$ N
Part b. Step 2. Again choose the left end of the diving board as the rotation point. Use the tip of pencil method to determine the direction of the torque.	Let the left end of the diving board be chosen as the rotation point. Pushing on the pencil at the point where F_B is located causes the pencil to rotate CCW. Thus F_B causes a counterclockwise torque about the about the left end of the board. Now repeat for the student's weight and the weight of the board. Note that each weight produces a clockwise torque (CW) about the left end of the board.
Part b. Step 3. Use the second condition of equilibrium to solve for the force exerted by support F_B.	$\Sigma\tau_{CCW} - \Sigma\tau_{CW} = 0$ F_B (1.0 m) sin 90° - (35 kg)(9.80 m/s^2)(2.0 m) sin 90° - (58 kg)(9.80 m/s^2)(4.0 m) sin 90° = 0 F_B (1.0 m) - 686 N m - 2270 N m = 0 $F_B = 3.0 \times 10^3$ N (upward)
Part b. Step 4. Apply the first condition of equilibrium, $\Sigma F = 0$, to solve for F_A.	From Part b. Step 1. $F_B - F_A = 9.1 \times 10^2$ N but $F_B = 3.0 \times 10^3$ N 3.0×10^3 N - $F_A = 9.1 \times 10^2$ N $F_A = 2.0 \times 10^3$ N (downward)

TEXTBOOK PROBLEM 24. The two trees in Fig. 9-58 are 7.6 m apart. A backpacker is trying to lift his pack out of the reach of bears. Calculate the magnitude of the force **F** that he must exert downward to hold a 19 kg backpack so that the rope sags at the midpoint by (a) 1.5 m, (b) 0.15 m.

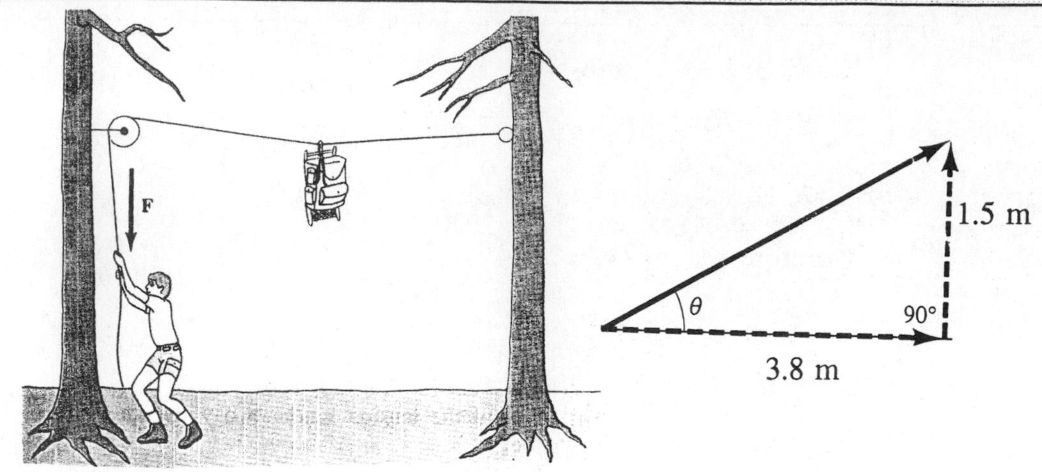

Part a. Step 1. Determine the angle θ that the rope makes with the horizontal.	Solution: (Sections 9-1, 9-2 and 9-3) Based on the diagram, the backpack is placed at the midpoint of the of the rope. Therefore, the angle θ between each section of the rope and the horizontal is the same. $\tan\theta = (1.5\text{ m})/(3.8\text{ m}) = 0.395$ $\theta = 21.5^\circ$
Part a. Step 2. Apply the first condition of equilibrium to the horizontal components of the tension in the rope.	There is no motion in the horizontal direction; therefore $\Sigma F_x = 0$ $T_{1x} - T_{2x} = 0$ $T_1 \cos\theta - T_2 \cos\theta = 0$ but the cos θ appears in each term. Therefore, $T_1 = T_2$.
Part a. Step 3. Apply the first condition of equilibrium to the vertical components of the forces acting on the rope. Note: the force exerted by the backpacker equals the tension in the rope.	There is no motion in the vertical direction; therefore, $\Sigma F_y = 0$. $T_{1y} + T_{2y} - mg = 0$ where mg is the weight of the backpack $T_1 \sin\theta + T_2 \sin\theta - mg = 0$ but $T_1 = T_2 = T$. Therefore simplifying, $2\,T \sin\theta = mg$ $T = (mg)/(2\sin\theta) = [(19\text{ kg})(9.8\text{ m/s}^2)]/[(2\sin 21.5^\circ)]$ $T = 250\text{ N}$

Part b. Step 1. Determine the angle θ the rope makes with the horizontal.	$\tan\theta = (0.15\ m)/(3.8\ m) = 0.0395$ $\theta = 2.26°$
Part b. Step 2. Apply the first condition of equilibrium to the vertical components of the forces acting on the rope. Note: The force exerted by the backpacker equals the tension in the rope.	Based on Part a. Step 1. $T_1 = T_2$. $T_{1y} + T_{2y} - mg = 0$ where mg is the weight of the backpack $T_1 \sin\theta + T_2 \sin\theta - mg = 0$ but $T_1 = T_2 = T$. Therefore simplifying, $2\ T \sin\theta = mg$ $T = (mg)/(2 \sin\theta) = [(19\ kg)(9.8\ m/s^2)]/[(2 \sin 2.26°)]$ $T = 2400\ N$

Note: As the backpack gets higher and higher, the value of θ becomes smaller. As θ becomes smaller, $\sin\theta$ becomes smaller and T becomes larger. As a result the force required to pull the backpack up increases as it gets higher above the ground. As θ approaches 0°, T approaches infinity. Even if the camper were able to pull with enough force to remove the sag, the rope would snap before the no sag point was reached.

TEXTBOOK PROBLEM 25. A door, 2.30 m high and 1.30 m wide, has a mass of 13.0 kg. A hinge 0.40 m from the top and another hinge 0.40 m from the bottom each support half of the door's weight (Fig. 9-59). Assume the center of gravity is at the geometrical center of the door, and determine the horizontal and vertical force components exerted by each hinge on the door.

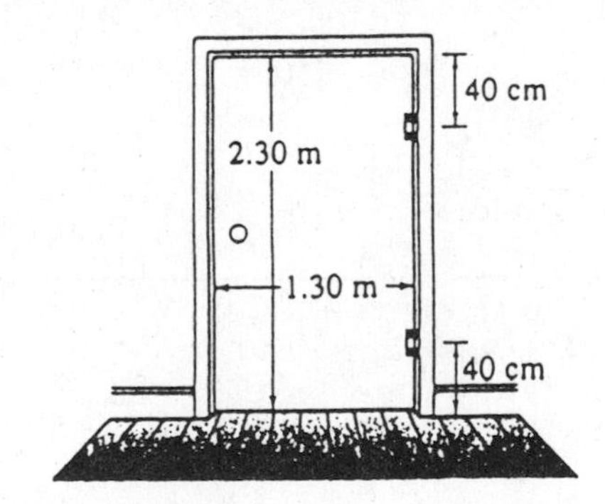

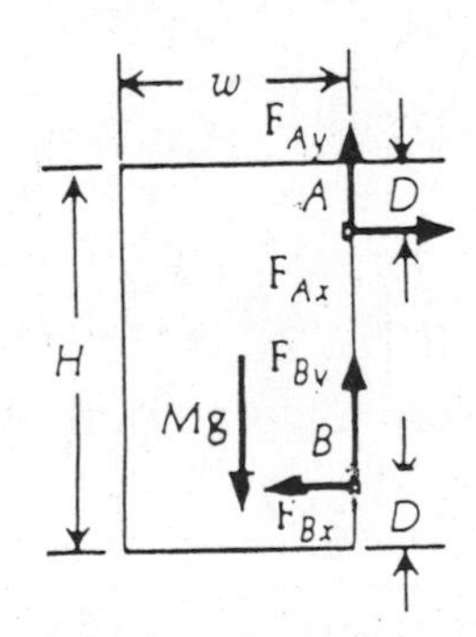

Part a. Step 1. Draw an accurate free body diagram locating the forces acting on the door. Determine the direction of the force exerted by each hinge.	Solution: (Sections 9-1, 9-2 and 9-3) Let F_B represent the force that the lower hinge exerts and F_A the force exerted by the upper hinge. In order to determine the direction of the force exerted by each hinge, let your textbook represent the door. Place one hand near the bottom edge of the book and the other hand near the top edge. If you release your upper hand the book rotates away from you. If you release your lower hand the book rotates toward you.

	As you can see, the forces exerted by the hinges are in the direction shown in the free body diagram.
Part a. Step 2. Apply the first condition of equilibrium, $\Sigma F = 0$.	$\Sigma F_x = 0 \quad F_{Ax} - F_{Bx} = 0$ $\Sigma F_y = 0 \quad F_{Ay} + F_{By} - (13.0\text{ kg})(9.80\text{ m/s}^2) = 0$ $F_{Ay} + F_{By} = 127\text{ N}$ As stated in the problem, each hinge supports half of the door's weight. Therefore, $F_{Ay} = F_{By} = 63.5\text{ N}$
Part a. Step 3. Use the tip of pencil method to determine the direction (CW or CCW) the of the torque produced by each force.	Since no rotation occurs, then any point may be selected as the location of the reference point. Therefore, let the lower hinge be chosen as the rotation point. Using point of pencil method, place the point of the pencil at the lower hinge with the length of the pencil parallel to the vertical height H of the door. Pushing on pencil at the point where F_{Ax} is located causes the pencil to rotate CW. Thus, F_{Ax} causes a clockwise torque about the top hinge. Again place the tip of the pencil at the lower hinge. Now place the pencil parallel to the bottom of the door and perpendicular to the line of action of the door's weight. The door's weight causes a CCW rotation about the lower hinge.
Part a. Step . 4 Use the second condition of equilibrium to solve for the force exerted by F_{Bx}.	$\Sigma\tau_{CCW} - \Sigma\tau_{CW} = 0$ $(13.0\text{ kg})(9.80\text{ m/s}^2)[(\frac{1}{2})(1.30\text{ m})(\sin 90^\circ)] - F_{Ax}[2.3\text{ m} - 2(0.40\text{ m})]\sin 90^\circ = 0$ $82.8\text{ Nm} - F_{Ax}[2.3\text{ m} - 2(0.40\text{ m})] = 0$ $F_{Ax} = (-82.8\text{ N m})/(-1.5\text{ m}) = 55.2\text{ N}$ (toward the right) From Part a. Step 2. $F_{Ax} - F_{Bx} = 0$ therefore $F_{Bx} = 55.2\text{ N}$ (toward the left)

TEXTBOOK PROBLEM 27. Consider a ladder with a painter climbing up it (Fig. 9-61). If the mass of the ladder is 12.0 kg, the mass of the painter is 55.0 kg, and the ladder begins to slip when her feet are 70% of the way up the length of the ladder, what is the coefficient of static friction between the ladder and the floor? Assume that the wall is frictionless.

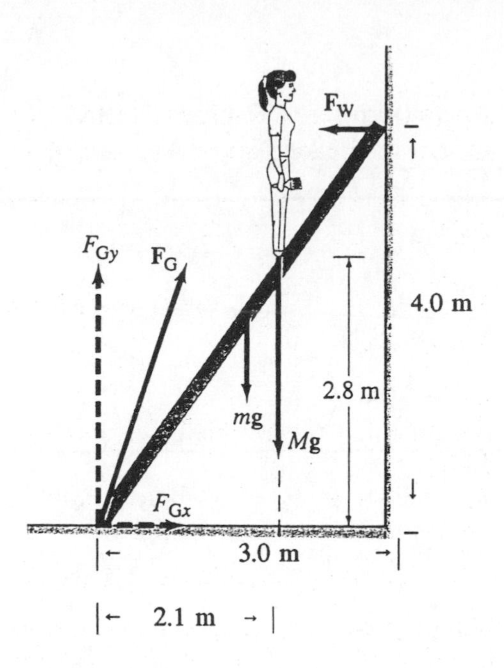

Part a. Step 1.

Determine the length of the ladder and the angle between the ground and the ladder.

Solution: (Sections 9-1, 9-2 and 9-3)

$L^2 = (3.0\ \text{m})^2 + (4.0\ \text{m})^2 = 25\ \text{m}^2$

$L = 5.0\ \text{m}$

$\tan\theta = (4.0\ \text{m})/(3.0\ \text{m}) = 1.33$

$\theta = \tan^{-1} 1.33 = 53^\circ$

Part a. Step 2.

Apply the first condition of equilibrium.

$\Sigma F_x = 0 \quad F_{Gx} - F_W = 0$

$\Sigma F_y = 0 \quad F_{Gy} - mg - Mg = 0$

$F_{Gy} - (12.0\ \text{kg})(9.8\ \text{m/s}^2) - (55.0\ \text{kg})(9.8\ \text{m/s}^2) = 0$

$F_{Gy} = 657\ \text{N}$

Part a. Step 3.

Apply the second condition of equilibrium, $\Sigma\tau = 0$.
Note: choose the point where the ladder rests on the ground as the rotation point.

The convenient point for the axis of rotation is a point where an unknown force is located. Therefore, either end of the ladder is a convenient point. Let us choose the end of the ladder that rests on the ground as the rotation point. At this point the ground exerts a force which is represented by F_{Gx} and F_{Gy}. Use the point of pencil method to determine the direction of each torque. From the diagram it can be seen that F_W produces a CCW torque while the weight of the ladder and the painter each produce a CW torque. Since F_{Gx} and F_{Gy} act at the rotation point, the lever arm distance is zero and they produce no torque.

$\Sigma\tau_{ccw} - \Sigma\tau_{cw} = 0$

$F_W\ (4.00\ \text{m})(\sin 90^\circ) - (12.0\ \text{kg})(9.8\ \text{m}^2)(2.50\ \text{m})(\cos 53^\circ)(\sin 90^\circ) - (55.0\ \text{kg})(9.8\ \text{m/s}^2)(0.70)(5.00\ \text{m})(\cos 53^\circ)(\sin 90^\circ) = 0$

	F_W (4.00 m) - 177 N m - 1135 N m = 0 F_W = 328 N
Part a. Step 4. Use Part a. Step 2. to determine F_{Gx}.	$F_{Gx} - F_W = 0$ F_{Gx} - 328 N = 0 F_{Gx} = 328 N
Part a. Step 5. Determine the coefficient of static friction between the ladder and the ground.	$F_{Gx} = F_{fr} = \mu_s F_N$ where $F_N = F_{Gy}$ = 657 N 328 N = μ_s (657 N) μ_s = 0.50

TEXTBOOK PROBLEM 39. A marble column of cross-sectional area 1.2 m^2 supports a mass of 25,000 kg. (a) What is the stress within the column? (b) What is the strain?

Part a. Step 1. Determine the stress.	Solution (Section: 9-5) stress = F/A = (25000 kg)(9.8 m/s^2)/(1.2 m^2) = 2.0 x 10^5 N/m^2
Part b. step 1. Determine the strain.	stress = (elastic modulus) strain $\Delta F/A = E (\Delta L/L_o)$ 2.0 x 10^5 N/m^2 = (50 x 10^9 N/m^2) strain strain = 4.1 x 10^{-6}

TEXTBOOK PROBLEM 72. A man doing push-ups pauses in the position shown in Fig. 9-84. His mass is 75 kg. Determine the normal force exerted by the floor (a) on each hand; (b) on each foot.

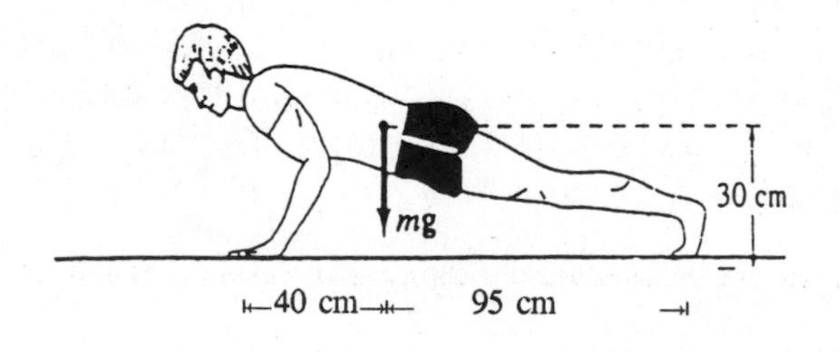

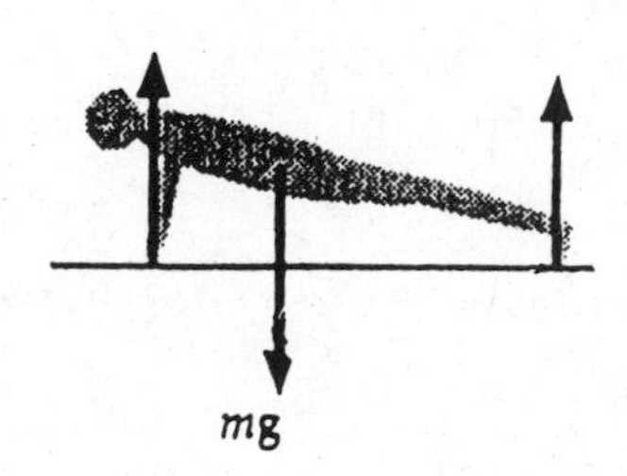

Part a. Step 1. Apply the first condition of equilibrium, $\Sigma F = 0$.	Solution: (Sections 9-2 and 9-3) $\Sigma F_x = 0$ No horizontal forces are present in this problem. $\Sigma F_y = 0$ $F_{hands} + F_{feet} - (75\ kg)(9.8\ m/s^2) = 0$ $F_{hands} + F_{feet} = 7.35 \times 10^2\ N$ (equation 1)
Part a. Step 2. Apply the second condition of equilibrium, The $\Sigma\tau = 0$.	It is convenient to select a point for the axis of rotation where an unknown force acts, e.g., the hands. Drawing a free body diagram and using the point of pencil method, we determine that the man's 7.35×10^2 N weight creates a CW torque about the hands. force produced by feet (F_{feet}) produces a CCW torque about the hands. The hands (F_{hands}) produce no torque since lever arm force distance to the rotation point equals zero. 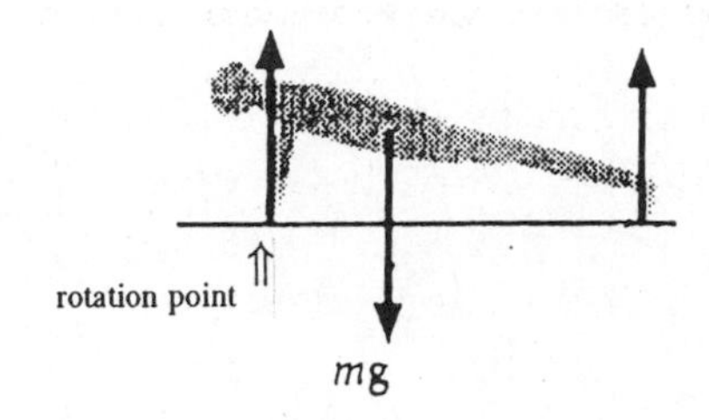 Applying the second condition of equilibrium: $\Sigma\tau_{CCW} - \Sigma\tau_{CW} = 0$ $F_{feet}(1.35\ m) - (7.35 \times 10^2\ N)(0.40\ m) = 0$ (equation 2) $F_{feet} = 2.2 \times 10^2\ N$ F_{feet} equals the total force exerted by the man's feet. Assuming that each foot supports half of the force, then each foot exerts a force of 1.1×10^2 N.
Part a. Step 4. Use equation 1 from Part a. Step 2. to solve for F_{hands}.	$F_{hands} + F_{feet} = 7.35 \times 10^2\ N$ $F_{hands} + 2.2 \times 10^2\ N = 7.35 \times 10^2\ N$ $F_{hands} = 5.2 \times 10^2\ N$ F_{hands} equals the total force exerted by the man's hands. Assuming that each hand supports half of the force, then each hand exerts a force of 2.6×10^2 N.

CHAPTER 10

FLUIDS

OBJECTIVES

After studying the material of this chapter, the student should be able to:

- distinguish between density, weight density, and specific gravity and given an object's mass and volume, calculate the object's density, weight density, and specific gravity.
- define pressure and calculate the pressure that an object of known weight exerts on a surface of known area, and express the magnitude of the pressure in psi, lb/ft^2, N/m^2, or pascals (Pa).
- calculate the pressure acting at a depth h below the surface of a liquid of density (ρ).
- distinguish between absolute pressure and gauge pressure and solve problems involving each type of pressure.
- state Pascal's principle and apply this principle to basic hydraulic systems.
- state Archimedes' principle and use this principle to solve problems related to buoyancy.
- explain what is meant by streamline flow, the equation of continuity, and the flow rate. Apply these concepts to word problems to solve for the velocity of water at a particular point in a closed pipe.
- use Bernoulli's equation and the concept of streamline flow to solve for the velocity of a fluid and/or the pressure exerted by a fluid at a particular point in a closed pipe.

KEY TERMS AND PHRASES

density (ρ) is the quantity of mass (m) per unit volume (V).

weight density is the ratio of the weight of a substance to its volume.

specific gravity (SG) is the ratio of the density of the substance to the density of pure water.

pressure is the ratio of the force acting perpendicular to a surface to the surface area (A) on which the force acts.

gauge pressure measures the difference in pressure between an unknown pressure and atmospheric pressure.

absolute pressure is the sum of the gauge pressure and the atmospheric pressure.

Pascal's principle states that pressure applied to a confined fluid is transmitted throughout the fluid and acts in all directions.

Archimedes' principle states that a body wholly or partly immersed in a fluid is buoyed up by a force equal to the weight of the fluid it displaces.

buoyant force is the force caused by the displaced fluid. The buoyant force is considered to be acting upward through the center of gravity of the displaced fluid.

streamline flow assumes that as each particle in the fluid passes a certain point it follows the same path as the particles that preceded it. There is no loss of energy due to internal friction in the fluid.

turbulent flow is the irregular movement of particles in a fluid and results in loss of energy due to internal friction in the fluid. Turbulent flow tends to increase as the velocity of a fluid increases.

flow rate is the mass (or volume) of fluid that passes a point per unit time.

Bernoulli's equation states that the quantity $P + \frac{1}{2} \rho v^2 + \rho g h$ is the same at every point in the moving fluid. This equation is derived through the use of the work-energy principle and ignores the effect of internal friction in a moving fluid. It should be noted that P, $\frac{1}{2} \rho v^2$, and $\rho g h$ all have dimensions of pressure.

SUMMARY OF MATHEMATICAL FORMULAS

density	$\rho = m/V$	Ratio of object's mass (m) to its volume (V).
specific gravity	$SG = \rho_{object} / \rho_{water}$	Ratio of the density of the substance to the density of pure water.
pressure	$P = F/A$	Ratio of the force acting perpendicular to a surface to the surface area (A) on which the force acts.
fluid pressure	$P = \rho g h$	Gauge pressure as a function of depth (h).
Archimedes' principle	$F_B = m_f g = \rho_f g V_f$	Buoyant force acting on an object placed in a fluid.
equation of continuity	$\rho A v = \text{constant}$	Mass of fluid moving through any point is constant.
volume flow rate	$R = v A$	Volume of fluid that passes a point per unit time.
Bernoulli's equation	$P + \frac{1}{2} \rho v^2 + \rho g h = \text{constant}$	Equation for the pressure at any point in a moving fluid.

CONCEPT SUMMARY

Density, Weight Density, and Specific Gravity

The **density** (ρ) of a substance is defined as the quantity of mass (m) per unit volume (V) and is given by

$\rho = m/V$

For solids and liquids, the density is usually expressed in grams per cm^3 (g/cm^3) or kilograms per cubic meter (kg/m^3). The density of gases is usually expressed in grams per liter (g/l).

The **weight density** of a substance is the ratio of the weight of a substance to its volume. weight density = weight/volume = $m\,g/V = \rho\,V\,g/V = \rho\,g$. Weight density is commonly used in problems involving English units. The weight density of distilled water at 39°F (4°C) is 62.4 lb/ft^3.

The **specific gravity** (SG) of a substance is the ratio of the density of the substance to the density of another substance which is taken as a standard. The density of pure water at 4°C is usually taken as the standard and this has been defined to be exactly 1.000 g/cm^3 or 1.000×10^3 kg/m^3.

Specific gravity is a dimensionless quantity. This is because it is the ratio of the density of one substance to the density of another. The specific gravity of an object is the same in any system of measurement.

SG = density of substance/density of water

For example, the density of aluminum is $2.7 \times 10^3\ kg/m^3$; therefore, the SG of aluminum is $SG = 2.7 \times 10^3\ kg/m^3 / 1.0 \times 10^3\ kg/m^3 = 2.7$.

EXAMPLE PROBLEM 1. A solid sphere made of wood has a radius of 0.100 m. The mass of the sphere is 1.0 kg. Determine the a) density and b) specific gravity of the wood.

Part a. Step 1. Determine the volume of the sphere.	Solution: (Section 10-1) $V = 4/3\ \pi\ r^3$ $V = 4/3\ \pi\ (0.100\ m)^3 = 4.18 \times 10^{-3}\ m^3$
Part a. Step 2. Apply the formula for density.	$\rho = m/V$ $\rho = (1.00\ kg)/(4.18 \times 10^{-3}\ m^3) = 239\ kg/m^3$
Part b. Step 1. Apply the formula for specific gravity.	SG = density of wood/density of water $SG = (239\ kg/m^3)/(1000\ kg/m^3) = 0.239$

Pressure

Pressure (P) is defined as the ratio of the force acting perpendicular to a surface to the surface area (A) over which the force acts.

$P = F/A$

The SI unit of pressure is N/m^2 or pascal (Pa) while the English unit is lb/ft^2. A number of other units are used, e.g., lb/in^2 (or psi) and atmospheres. One atmosphere is the average pressure exerted by the Earth's atmosphere at sea level.

$1.00 \text{ atm} = 14.7 \text{ lb/in}^2 = 1.01 \times 10^5 \text{ N/m}^2 = 101.3 \text{ kPa}$

TEXTBOOK QUESTION 5. A small amount of water is boiled in a 1-gallon metal can. The can is removed from the heat and the lid put on. Shortly thereafter the can collapses. Explain.

ANSWER: When the lid is placed on the can, the internal pressure exerted by the hot gases, water vapor, and air inside the can balances the external atmospheric pressure acting on the can. As the gases inside the can cool, the pressure they exert on the side walls of the can decreases while the external atmospheric pressure remains constant. Eventually, the difference in pressure becomes large enough that the can collapses.

Pressure in Fluids

The pressure (P) produced by a column of fluid of height h and density ρ is given by

$$P = \rho g h$$

The density of liquids and solids is considered to be constant. In reality, the density of a liquid will increase slightly with increasing depth. The variation in density is usually negligible and can be ignored.

Gauge Pressure and Absolute Pressure

Ordinary pressure gauges measure the difference in pressure between an unknown pressure and atmospheric pressure. The pressure measured is called the **gauge pressure** and the unknown pressure is referred to as the **absolute pressure**.

$$P_{absolute} = P_{gauge} + P_{atmosphere}$$

Thus, if a tire gauge registers 193 kPa, the absolute pressure would be approximately 193 kPa + 101 kPa = 294 kPa.

EXAMPLE PROBLEM 2. A submerged wreck is located 18.3 m (i.e., 60 feet) beneath the surface of the ocean off the coast of South Florida. Determine the a) gauge pressure and b) absolute pressure on a scuba diver who is exploring the wreck. Note: the density of seawater is 1025 kg/m^3.

Part a. Step 1. Calculate the gauge pressure at 18.3 m.	$P = \rho\, g\, h$ $P = (1025\ kg/m^3)(9.80\ m/s^2)(18.3\ m)$ $P = 1.83 \times 10^5\ N/m^2$ or 182 kPa
Part b. Step 1. Calculate the absolute pressure on the diver.	$P_{absolute} = P_{gauge} + P_{atmosphere}$ where $1.0\ atm = 1.013 \times 10^5\ N/m^2$ $= 1.83 \times 10^5\ N/m^2 + 1.013 \times 10^5\ N/m^2$ $P_{absolute} = 2.84 \times 10^5\ N/m^2$ or 284 kPa

Pascal's Principle

Pascal's principle states that pressure applied to a confined fluid is transmitted throughout the fluid and acts in all directions. The principle means that if the pressure on any part of a confined fluid is changed, then the pressure on every other part of the fluid must be changed by the same amount. This principle is basic to all hydraulic systems.

EXAMPLE PROBLEM 3. In the hydraulic system shown in the diagram, the 200 kg cylinder has a cross-sectional area of 50.0 cm^2. The cylinder on the right has a cross-sectional area of 2.50 cm^2. a) Determine the weight (F) required to hold the system in equilibrium. b) If the left-hand cylinder is pushed down 10.0 cm, determine through what distance F will move.

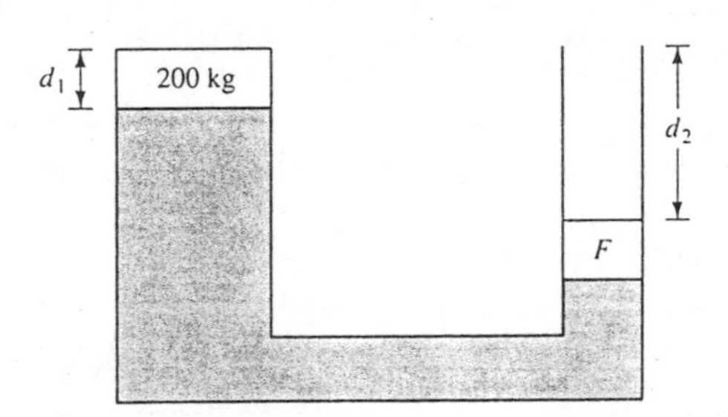

Part a. Step 1. Apply Pascal's principle and solve for F.	Solution: (Section 10-4) $(F_1)/(A_1) = (F_2)/(A_2)$ $[(200\ kg)(9.80\ m/s^2)]/(50.0\ cm^2) = (F_2)/(2.50\ cm^2)$ $F = 98.0\ N$
Part b. Step 1. Determine the distance that the right-hand cylinder rises.	The volume of fluid displaced by the 50.0 cm^2 cylinder equals the change in the volume of fluid in the right-hand cylinder. Note: the volume of the displaced fluid equals the product of the cross-sectional area of the piston and the distance that the piston moves. $V = A\, d$ $A_1\, d_1 = A_2\, d_2$

$(50.0 \text{ cm}^2)(10.0 \text{ cm}) = (2.50 \text{ cm}^2)\, d_2$

$d_2 = 2.00 \times 10^2 \text{ cm}$

Buoyancy and Archimedes' Principle

Archimedes' principle states that a body wholly or partly immersed in a fluid is buoyed up by a force equal to the weight of the fluid it displaces. This upward force is called the **buoyant force**. As a result, objects lowered into a fluid "appear" to lose weight. The buoyant force is considered to be acting upward through the center of gravity of the displaced fluid.

TEXTBOOK QUESTION 14. Explain why helium weather balloons, which are used to measure atmospheric conditions at high altitudes, are normally released while filled to only 10%-20% of their maximum volume.

ANSWER. A weather balloon will rise as long as the buoyant force exerted by the air is slightly greater than the balloon's weight. As the balloon rises the external air pressure will decrease and the balloon's volume will increase as the inside pressure matches the outside pressure. If the balloon were to be fully inflated at ground level, the balloon would prematurely expand at a lower altitude, the material would be overstretched and it would burst and fall back to the ground. If the balloon is only partially inflated, the balloon's volume would gradually increase as it rises and the balloon would reach a much higher altitude. At that point, the balloon's weight equals the buoyant force. Eventually, helium molecules will diffuse out of the balloon and the balloon will gradually descend to the ground.

EXAMPLE PROBLEM 4. An aluminum object has a mass of 13.5 kg and a density of 2.70×10^3 kg/m^3. The object is attached to a string and immersed in a tank of water. Determine the a) volume of the object and b) tension in the string when it is completely immersed.

Part a. Step 1.

Determine the volume of the object.

Solution: (Sections 10-1 and 10-6)

$\rho = m/V$ and

$V = m/\rho = 13.5 \text{ kg}/2.70 \times 10^3 \text{ kg/m}^3 = 5.00 \times 10^{-3} \text{ m}^3$

Part b. Step 1.

Draw an accurate diagram and locate all of the forces acting on the object.

T is the tension in the string, mg is the object's weight, and the buoyant force is F_b.

Part b. Step 2. Determine the volume of water displaced by the object.	Since the object is completely submerged, the volume of the water displaced equals the volume of the aluminum object. $V_{water} = 5.00 \times 10^{-3}\ m^3$
Part b. Step 3. Determine the weight of the water displaced by the submerged object.	The weight of the water displaced can be determined since both the density and the volume of the water displaced are known. $weight_{water\ displaced} = \rho_w V g$ $= (1.00 \times 10^3\ kg/m^3)(5.00 \times 10^{-3}\ m^3)(9.80\ m/s^2)$ $weight_{water\ displaced} = 49.0\ N$
Part b. Step 4. Use Archimedes' principle to determine the buoyant force.	Archimedes' principle states that the buoyant force equals the weight of the fluid displaced; therefore, $F_B = 49.0\ N$
Part b. Step 5. Apply Newton's second law and solve for the tension in the string.	$\Sigma F = m\ a$ but $a = 0\ m/s^2$ $T + F_B - mg = 0$ $T + 49.0\ N - (13.5\ kg)(9.80\ m/s^2) = 0$ $T = 83.3\ N$

Fluids in Motion

The equations that follow are applied when a moving fluid exhibits **streamline flow**. Streamline flow assumes that as each particle in the fluid passes a certain point it follows the same path as the particles that preceded it. There is no loss of energy due to internal friction in the fluid.

In reality, particles in a fluid exhibit **turbulent flow**, which is the irregular movement of particles in a fluid and results in loss of energy due to internal friction in the fluid. Turbulent flow tends to increase as the velocity of a fluid increases.

Flow Rate

The **flow rate** is the mass of fluid that passes a point per unit time (m/t). The formula for flow rate is

$$m/t = \rho\ A\ v$$

where ρ is the density of the fluid, A is the cross-sectional area of the tube of fluid at the particular point in question, and v is the velocity of the fluid at the point in question.

Since fluid cannot accumulate at any point, the flow rate is constant. This is expressed as the **equation of continuity**.

$\rho A v = \text{constant}$

In streamline flow, the fluid is considered to be incompressible and the density is the same throughout. The equation of continuity can then be written in terms of the volume rate of flow (R) which is constant throughout the fluid: $R = Av = \text{constant}$.

Bernoulli's Equation

A fluid in streamline flow follows **Bernoulli's equation**:

$P + \frac{1}{2} \rho v^2 + \rho g h = \text{constant}$

P, ρ, and h are the absolute pressure, density, and height at a particular point in a fluid. The equation states that the quantity $P + \frac{1}{2} \rho v^2 + \rho g h$ is the same at every point in the stream. This equation is derived through the use of the work-energy principle and ignores the effect of internal friction in a moving fluid. It should be noted that P, $\frac{1}{2} \rho v^2$, and ρgh all have dimensions of pressure.

TEXTBOOK QUESTION 17. If you dangle two pieces of paper vertically, a few inches apart (Fig. 10-46) and blow air between them, how do you think the papers will move? Try it and see. Explain.

ANSWER: Initially the papers move toward each other. Moving air between the pieces causes a decrease in the air pressure between the pieces. The pressure outside the gap is greater than the air pressure on the inside forcing the pieces together.

The air flow momentarily stops when the pieces of paper meet. At this point, there is no difference in the air pressure between the pieces as compared to outside the pieces and the pieces begin to move apart. However, as the pieces move apart air again flows between the pieces and the pieces again move together. The result is that the pieces tend to flutter.

TEXTBOOK QUESTION 18. Why does the canvas top of a convertible bulge out when the car is traveling at high speed?

ANSWER: Air moving over the top of the canvas causes a decrease in the downward pressure (P_2). Inside the car the speed of the air is much less and therefore the upward pressure (P_1) on the canvas is greater than the downward pressure. The pressure difference ($P_1 - P_2$) causes the canvas top to bulge out. Using Bernoulli's equation, and setting point 2 above the canvas while point 1 is a point below the canvas, then

$P_1 + \frac{1}{2} \rho v_1^2 + \rho g h_1 = P_2 + \frac{1}{2} \rho v_2^2 + \rho g h_2$

but $h_1 \approx h_2$ and $v_1 = 0$. Then $P_1 - P_2 = \frac{1}{2} \rho v_2^2$.

As the car's speed increases, the difference in pressure increases and the canvas bulges out.

EXAMPLE PROBLEM 5. During the early morning hours of August 24, 1992, the winds of Hurricane Andrew visited the home of the author of this study guide. It is estimated that wind gusts in excess of 150 mph (67 m/s) passed over the roof of the author's house. Calculate the a) difference in pressure between the air in the attic and the air passing over the roof and b) upward force exerted on the roof if area of the roof is 275 m^2. Assume that the air inside the attic was not moving and that the thickness of the roof was negligible. Note: assume that the density of air is 1.29 kg/m^3.

Part a. Step 1.	Solution: (Section 10-8)
Use Bernoulli's equation to determine the pressure.	$P_1 + \frac{1}{2} \rho v_1^2 + \rho g h_1 = P_2 + \frac{1}{2} \rho v_2^2 + \rho g h_2$ But $h_1 = h_2$ while $v_1 = 0$. Therefore, the equation becomes
Let P_1 represent the pressure inside the attic and $v_1 = 0$. Also, since the thickness of the roof is negligible $h_1 = h_2$.	$P_1 = P_2 + \frac{1}{2} \rho v_2^2$ $P_1 - P_2 = \frac{1}{2} \rho v_2^2 = \frac{1}{2} (1.29 \text{ kg/m}^3)(67 \text{ m/s})^2$ $P_1 - P_2 = 2.9 \times 10^3 \text{ N/m}^2$
Part b. Step 1.	pressure = force/area and force = pressure x area
Calculate the upward force (lift) exerted by the moving air on the roof.	$F = (2.9 \times 10^3 \text{ N/m}^2)(275 \text{ m}^2) = 8.0 \times 10^5 \text{ N}$ Note: this upward force is approximately equal to 90 tons of lift. According to the Miami-Dade County building code all roof trusses must be attached to the tie beam of the building by metal straps. The tile on the author's roof needed to be replaced but the roof did stay on the house.

EXAMPLE PROBLEM 6. A horizontal pipe has a diameter of 0.150 m at point 1 and 0.0500 m at point 2. The velocity of water at point 1 is 0.800 m/s and the pressure is 1.01 x 10^5 N/m^2. Determine the a) volume flow rate, b) velocity of the water at point 2, and c) pressure at point 2.

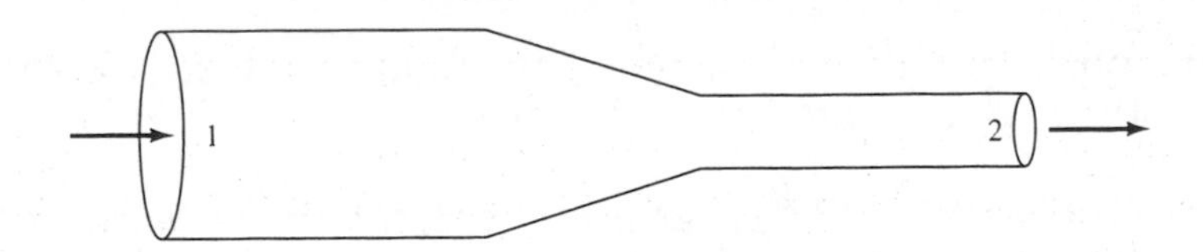

Part a. Step 1.	Solution: (Sections 10-7 and 10-8)
Calculate the volume flow rate.	$R = A_1 v_1 = [\pi(0.150 \text{ m}/2)^2](0.800 \text{ m/s}) = 1.41 \times 10^{-2} \text{ m}^3/\text{s}$

Part b. Step 1. Apply the rate flow equation and determine the velocity of the water at point 2.	Water cannot accumulate at any point in the hose, the rate of volume flow (R) must be the same throughout. $R = A_1 v_1 = A_2 v_2$ where $A = \pi(\text{diameter}/2)^2$ $[\pi(0.150 \text{ m}/2)^2](0.800 \text{ m/s}) = [\pi(0.0500 \text{ m}/2)^2] v_2$ $v_2 = 7.20 \text{ m/s}$
Part c. Step 1. Use Bernoulli's equation to determine the pressure. Note: the pipe is horizontal; therefore, $h_1 = h_2$.	$P_1 + ½ \rho v_1^2 + \rho g h_1 = P_2 + ½ \rho v_2^2 + \rho g h_2$ but $h_1 = h_2$, $P_1 + ½ \rho v_1^2 = P_2 + ½ \rho v_2^2$ $1.01 \times 10^5 \text{ N/m}^2 + ½ (1000 \text{ kg/m}^3)(0.800 \text{ m/s})^2 =$ $P_2 + ½ (1000 \text{ kg/m}^3)(7.20 \text{ m/s})^2$ $1.01 \times 10^5 \text{ N/m}^2 + 320 \text{ N/m}^2 = P_2 + 2.59 \times 10^4 \text{ N/m}^2$ $P_2 = 7.54 \times 10^4 \text{ N/m}^2$

PROBLEM SOLVING SKILLS

For problems involving density and specific gravity:

1. If necessary, use information given in the problem to calculate the object's mass and volume.
2. Determine the object's density ($\rho = m/V$).
3. Determine the object's specific gravity by dividing the object's density by the density of water.

For problems involving Pascal's principle:

1. Identify which quantities are related to the input and which relate to the output and complete a data table based on information given and implied in the problem.
2. Use the concept of pressure in a fluid and Pascal's principle to solve the problem.

For problems where Archimedes' principle is used to calculate the buoyant force on an object completely immersed in a fluid:

1. If the mass and density of the object are known, then the volume of the object can be determined.
2. The volume of the object equals the volume of the fluid the object displaces.
3. Use Archimedes' principle to determine the buoyant force.

For problems where Archimedes' principle is used to calculate the buoyant force on an object that floats in the fluid:

1. For a floating object the buoyant force equals the object's weight.
2. The weight of the fluid displaced equals the buoyant force.

For problems involving the rate flow equation and the equation of continuity:

1. Complete a data table listing the cross-sectional area of the closed pipe at each point, as well as the velocity and density of the fluid at the each point in question.

2. Use the rate flow equation and equation of continuity to solve the problem.

For problems involving Bernoulli's equation:

1. Complete a data table listing the pressure, velocity of the fluid, and height of the fluid above a reference point for each point in question. If the problem involves a fluid moving through a closed pipe, then determine the cross-sectional area of the closed pipe at each point in question.
2. If necessary, use the equation of continuity to determine the velocity of the fluid at a particular point.
3. Use Bernoulli's equation to solve the problem.

SOLUTIONS TO SELECTED TEXTBOOK PROBLEMS

TEXTBOOK PROBLEM 3. If you tried to smuggle gold bricks by filling your backpack, whose dimensions are 60 cm x 28 cm x 18 cm, what would its mass be?

Part a. Step 1. Determine the volume of the gold in cubic meters (m^3).	Solution: (Section 10-2) Volume = (length)(width)(height) $V = (0.60\ m)(0.28\ m)(0.18\ m)$ $V = 3.0 \times 10^{-2}\ m^3$
Part a. Step 2. Determine the gold's mass. Note: the density of gold is $19.3 \times 10^3\ kg/m^3$.	density = mass/volume mass = (density)(volume) mass = $(19.3 \times 10^3\ kg/m^3)(3.0 \times 10^{-2}\ m^3) = 5.8 \times 10^2\ kg \approx 1300$ lbs.

TEXTBOOK PROBLEM 7. Estimate the pressure exerted on a floor by (a) a pointed chair leg (60 kg on all four legs) of area = 0.020 cm^2, and (b) a 1500 kg elephant standing on one foot (area = 800 cm^2).

Part a. Step 1. Convert the area from cm^2 to m^2.	Solution: (Section 10-3) chair $(4\ legs)(0.020\ cm^2)(1\ m^2/10^4\ cm^2) = 8.0 \times 10^{-6}\ m^2$ elephant $(800\ cm^2)(1\ m^2/10^4\ cm^2) = 8.0 \times 10^{-2}\ m^2$
Part a. Step 2. Determine the pressure exerted by one chair leg and the elephant.	$P_{leg} = (F_{chair})/(A_{legs})$ $= [(60\ kg)(9.8\ m/s^2)]/(8.0 \times 10^{-6}\ m^2)$ $P_{leg} \approx 7 \times 10^7\ N/m^2$ $P_{elephant} = (F_{elephant})/(A_{elephant})$

	$= [(1500 \text{ kg})(9.8 \text{ m/s}^2)]/(8.0 \times 10^{-2} \text{ m}^2)$ $P_{elephant} = 1.8 \times 10^5 \text{ N/m}^2 \approx 2 \times 10^5 \text{ N/m}^2$
Part a. Step 3. Determine the ratio of the pressures.	$P_{chair\ leg}/P_{elephant} \approx (7 \times 10^7 \text{ N/m}^2)/(2 \times 10^5 \text{ N/m}^2) \approx 400$ One chair leg exerts a pressure approximately 400 times greater than that of the elephant placing its weight on one foot.

TEXTBOOK PROBLEM 33. The specific gravity of ice is 0.917, whereas for seawater is 1.025. What fraction of an iceberg is above the water?

Part a. Step 1. Determine the density of ice and seawater.	Solution: (Section 10-6) $SG = (\rho_{ice})/(\rho_{water})$ $\rho_{ice} = (SG_{ice})(\rho_{water}) = (0.917)(1000 \text{ kg/m}^3) = 917 \text{ kg/m}^3$ $SG = (\rho_{seawater})/(\rho_{water})$ $\rho_{seawater} = (SG_{seawater})(\rho_{water}) = (1.025)(1000 \text{ kg/m}^3) = 1025 \text{ kg/m}^3$
Part a. Step 2. Use Archimedes determine the fraction of an iceberg's volume is above the water.	According to Archimedes' principle, the buoyant force equals the weight of the water displaced. Since iceberg floats, the buoyant force equals the weight of the iceberg. $F_B = m_{seawater}\, g$ and $F_B = m_{iceberg}\, g$ $m_{seawater}\, g = m_{iceberg}\, g$ Note: g cancels $m_{seawater} = m_{iceberg}$ and $m = \rho V$ $\rho_{seawater} V_{seawater} = \rho_{iceberg} V_{iceberg}$ $(1025 \text{ kg/m}^3)V_{seawater} = (917 \text{ kg/m}^3)V_{iceberg}$ $V_{seawater} = [(917 \text{ kg/m}^3)V_{iceberg}]/(1025 \text{ kg/m}^3)$ $V_{seawater} = 0.895\, V_{iceberg}$ The amount of water displaced by the iceberg is 0.895, i.e. 89.5%, of the iceberg's volume. Therefore, 89.5% of the iceberg's volume is below the ocean's surface while 0.105 or 10.5% is visible above the surface.

TEXTBOOK PROBLEM 44. What is the lift (in newtons) due to Bernoulli's principle on a wing of area 78 m^2 if the air passes over the top and bottom surfaces at speeds of 260 m/s and 150 m/s, respectively.

Part a. Step 1. Complete a data table using information both given and implied.	Solution: (Section 10-8) Let P_1 represent the pressure below the wing while P_2 represents the pressure above the wing. Also, the thickness of the wing is negligible $h_1 \approx h_2$. $P_1 = ?$ $P_2 = ?$ $v_1 = 360$ m/s $v_2 = 150$ m/s $\rho = 1.29$ kg/m^3 $h_1 \approx h_2$
Part a. Step 2. Use the Bernoulli equation to determine the difference in pressure.	$P_1 + \frac{1}{2} \rho v_1^2 + \rho g h_1 = P_2 + \frac{1}{2} \rho v_2^2 + \rho g h_2$ Rearranging the equation gives $P_1 - P_2 = \frac{1}{2} \rho (v_2^2 - v_1^2) + \rho g (h_2 - h_1)$ but $\rho g (h_2 - h_1) \approx 0$ $P_1 - P_2 = \frac{1}{2} (1.29 \text{ kg/m}^3)[(260 \text{ m/s})^2 - (150 \text{ m/s})^2]$ $P_1 - P_2 = 2.9 \times 10^4 \text{ N/m}^2$
Part b. Step 1. Calculate the upward force (lift) exerted by the moving air on the wing.	pressure = force/area and force = pressure x area $F = (2.9 \times 10^4 \text{ N/m}^2)(78 \text{ m}^2) = 2.3 \times 10^6$ N Note: this upward force is approximately equal to 258 tons of lift.

TEXTBOOK PROBLEM 46. Water at a gauge pressure of 3.8 atm at street level flows into an office building at a speed of 0.60 m/s through a pipe 5.0 cm in diameter. The pipe tapers down to 2.6 cm in diameter by the top floor, 18 m above (Fig. 10-53), where the faucet has been left open. Calculate the flow velocity and the gauge pressure in such a pipe at the top floor. Assume no branch pipes and ignore viscosity.

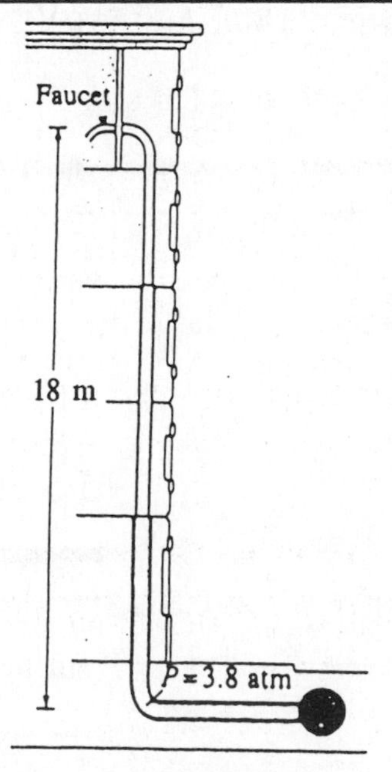

Part a. Step 1. Complete a data table using information both given and implied.	Solution: (Sections 10-8 and 10-9) Assume that point 1 is at $h_1 = 0$ and that $h_2 = 18$ m. $P_1 = (3.8 \text{ atm})[(1.01 \times 10^5 \text{ N/m}^2)/(1 \text{ atm}) = 3.84 \times 10^5 \text{ N/m}^2$ $\rho_{water} = 1.0 \times 10^3 \text{ kg/m}^3$ $g = 9.8 \text{ m/s}^2$ $v_1 = 0.60 \text{ m/s}$ $h_1 = 0$ m $A_1 = \pi(0.050 \text{ m})^2 = 7.8 \times 10^{-3} \text{ m}^2$ $A_2 = \pi(0.026 \text{ m})^2 = 2.1 \times 10^{-3} \text{ m}^2$ $h_2 = 18$ m $v_2 = ?$ $P_2 = ?$
Part a. Step 2. Use the rate flow equation to solve for the velocity at point 2.	The volume rate flow remains constant throughout the pipe. Determine v_2 using the rate flow equation. $A_1 v_1 = A_2 v_2$ $(7.8 \times 10^{-3} \text{ m}^2)(0.60 \text{ m/s}) = (2.1 \times 10^{-3} \text{ m}^2) v_2$ $v_2 = 2.2$ m/s
Part a. Step 3. Use Bernoulli's equation to determine the gauge pressure at point 2.	$P_1 + \frac{1}{2} \rho v_1^2 + \rho g h_1 = P_2 + \frac{1}{2} \rho v_2^2 + \rho g h_2$ but $h_1 = 0$, then $\rho g h_1 = 0$ $P_1 + \frac{1}{2} \rho v_1^2 = P_2 + \frac{1}{2} \rho v_2^2 + \rho g h_2$ $3.8 \times 10^5 \text{ N/m}^2 + \frac{1}{2} (1000 \text{ kg/m}^3)(0.60 \text{ m/s})^2 =$ $P_2 + \frac{1}{2} (1000 \text{ kg/m}^3)(2.2 \text{ m/s})^2 + (1000 \text{ kg/m}^3)(9.8 \text{ m/s}^2)(18 \text{ m})$ $3.8 \times 10^5 \text{ N/m}^2 + 180 \text{ N/m}^2 = P_2 + 2.4 \times 10^3 \text{ N/m}^2 + 1.76 \times 10^5 \text{ N/m}^2$ $P_2 = 2.0 \times 10^5 \text{ N/m}^2 \approx 2.0$ atm

TEXTBOOK PROBLEM 78. A raft is made of 10 logs lashed together. Each is 56 cm in diameter and has a length of 6.1 m. How many people can the raft hold before they start getting their feet wet, assuming that the average person has a mass of 68 kg? Do not neglect the weight of the logs. Assume that the specific gravity of wood is 0.60.

Part a. Step 1. Determine the density of the wood.	Solution: (Section 10-6) $SG = (\rho_{wood})/(\rho_{water})$ $\rho_{wood} = (SG)(\rho_{water}) = (0.60)(1000 \text{ kg/m}^3) = 600 \text{ kg/m}^3$

Part a. Step 2. Calculate the total volume of the raft and the volume of the water displaced.	The raft is made of cylindrical logs. The radius of a log = ½ (56 cm) = 28 cm = 0.28 m. The volume of a log given by $V = \pi r^2 h = \pi (0.28 \text{ m})^2(6.1 \text{ m}) = 1.5 \text{ m}^3$ The raft is made up of 10 logs. The total volume of the raft is $V_{raft} = (10)(1.5 \text{ m}^3) = 15 \text{ m}^3$ According to the statement of the problem, the raft is "just" submerged. Therefore, the volume of the water displaced equals the volume of the raft, i.e., $V_{water} = 15 \text{ m}^3$
Part a. Step 3. Use Archimedes' principle to determine the maximum number of people who can stand on the raft without getting their feet wet.	According to Archimedes' principle, the buoyant force equals the weight of the water displaced. The top of the raft is just at the water line, the buoyant force of the water equals the weight of the raft + people. $F_B = m_{water}\, g$ and $F_B = m_{raft}\, g + m_{people}\, g$ $m_{water}\, g = m_{raft}\, g + m_{people}\, g$ Note: g cancels $m_{water} = m_{raft} + m_{people}$ and $m = \rho V$ $\rho_{water} V_{water} = \rho_{wood} V_{wood} + m_{people}$ $(1000 \text{ kg/m}^3)(15 \text{ m}^3) = (600 \text{ kg/m}^3)(15 \text{ m}^3) + m_{people}$ $15000 \text{ kg} = 9000 \text{ kg} + m_{people}$ $m_{people} = 6000 \text{ kg}$ Each person has a mass of 68 kg; therefore, maximum number of people = (6000 kg)/(68 kg) $\approx$ 88

CHAPTER 11

VIBRATIONS AND WAVES

OBJECTIVES

After studying the material of this chapter, the student should be able to:

- state the conditions required to produce SHM.
- determine the period of motion of an object of mass m attached to a spring of force constant k.
- calculate the velocity, acceleration, potential, and kinetic energy at any point in the motion of an object undergoing SHM.
- write equations for displacement, velocity, and acceleration as sinusoidal functions of time for an object undergoing SHM if the amplitude and angular velocity of the motion are known. Use these equations to determine the displacement, velocity, and acceleration at a particular moment of time.
- determine the period of a simple pendulum of length L.
- state the conditions necessary for resonance. Give examples of instances where resonance is a) beneficial and b) destructive.
- explain how damped harmonic motion can be achieved to prevent destructive resonance.
- distinguish between a longitudinal wave and a transverse wave and give examples of each type of wave.
- calculate the speed of longitudinal waves through liquids and solids and the speed of transverse waves in ropes and strings.
- calculate the energy transmitted by a wave, the power of a wave and the intensity of a wave, across a unit area A.
- describe wave reflection from a barrier, refraction as the wave travels from one medium into another, constructive and destructive interference as waves overlap, and diffraction of waves as they pass around an obstacle.
- explain how a standing wave can be produced in a string or rope and calculate the harmonic frequencies needed to produce standing waves in string instruments.

KEY TERMS AND PHRASES

simple harmonic motion (SHM) refers to periodic vibrations or oscillations that exhibit two characteristics: 1) the force acting on the object and the magnitude of the object's acceleration are always directly proportional to the displacement of the object from its equilibrium position, and 2) both the force vector and the acceleration vector are directed opposite to the displacement vector and therefore in toward the object's equilibrium position.

amplitude (A) of an object undergoing SHM refers to the maximum displacement of the object from the equilibrium position.

period (T) of the motion is the time required for the motion to repeat.

frequency (f) refers to the number of complete repetitions of the motion that occur each second. The frequency is inversely related to the period.

simple pendulum is assumed to have its entire mass concentrated at the end of its length. The simple pendulum undergoes SHM if the maximum angle that it is displaced from equilibrium is small (approximately 15° or less).

damping is the loss of mechanical energy as the amplitude of motion in a simple harmonic oscillator gradually decreases.

forced vibrations have the same frequency as the external force and not necessarily equal to the natural frequency of the object.

resonance occurs if the external forced vibrations have the same frequency as one of the natural frequencies of the object. The natural frequency (or frequencies) at which resonance occurs is called the **resonant frequency**.

transverse wave is a wave in which the particles of the medium move at right angles to the direction of motion of the wave.

crest is the highest point of that portion of a transverse wave above the equilibrium position.

trough is the lowest point of that portion of a transverse wave below the equilibrium position.

longitudinal wave is a wave in which the particles of the medium move back and forth, parallel to the direction of the motion of the wave.

compressions or **condensations** are regions in a longitudinal wave where the density of particles of the medium is greater than when the medium is at equilibrium. It is convenient to compare a region of compression with the crest of a transverse wave.

expansions or **rarefactions** are regions in a longitudinal wave where the density of particles of the medium is less than when the medium is at equilibrium. It is convenient to compare an expansion with the trough of a transverse wave.

wavelength (λ) is the distance between any two repeating points on a periodic wave, e.g., the distance between adjacent crests or adjacent compressions.

intensity (I) of a wave is defined as the power transmitted across a unit area (A) perpendicular to the direction of energy flow.

interference results when two waves pass through the same region of space at the same time.

principle of superposition states that when two waves pass through a medium at the same time, the resultant displacement of the medium at any particular moment of time equals the algebraic sum of the displacements of the component waves at that point.

destructive interference occurs if the amplitude of the resultant of two interfering waves is smaller than the displacement of either wave.

constructive interference occurs if the amplitude of the resultant of two interfering waves is larger than the displacement of either wave.

diffraction refers to the ability of waves to bend around obstacles. The amount of diffraction depends on the wavelength of the waves and the size of the obstacle.

standing waves are produced by the superposition of two periodic waves having identical frequencies and amplitudes which are traveling in opposite directions.

nodal points are fixed positions along the entire length of a standing wave where the displacement is always zero. The nodal points are found at half wavelength ($\frac{1}{2}\lambda$) intervals along the length of the medium.

antinodal points are points of maximum amplitude located halfway between adjacent nodal points. The displacement of the medium at the point fluctuates between a crest and a trough. The amplitude of the wave at the antinodal points equals the sum of the amplitudes of the two component waves which produce the standing wave pattern.

first harmonic or **fundamental frequency** or **first mode of vibration** refers to the lowest possible frequency that can produce a standing wave. Overtones refer to higher frequencies which also produce standing waves.

SUMMARY OF MATHEMATICAL FORMULAS

period of an object in SHM at the end of a spring	$T = 2\pi\,(m/k)^{1/2}$	The period (T) is related to the object's mass (m) as well as the force constant (k) of the spring.
velocity of an object in SHM at the end of a spring	$v = v_o\,(1 - x^2/A^2)^{1/2}$ or $v = [(k/m)(A^2 - x^2)]^{1/2}$	v is the velocity of the object at displacement x from equilibrium and v_o is the maximum velocity of the object. A is the amplitude of the motion.
total energy of an object in SHM at the end of a spring	$E = \frac{1}{2}\,m\,v^2 + \frac{1}{2}\,k\,x^2$	The total energy (E) equals the sum of the kinetic energy and the potential energy.
period of a simple pendulum	$T = 2\pi(L/g)^{1/2}$	The period of a simple pendulum depends on its length and the acceleration due to gravity.
wave speed	$v = f\,\lambda$	The speed (v) of a periodic wave equals the product of the frequency (f) and the wavelength (λ).

displacement of a particle in SHM	$x = A \cos \omega t$ $\omega = 2 \pi f = 2\pi/T$	The displacement (x) of a particle in a transverse wave in SHM as a function of time t. Note: at $t = 0$, $x = A$.
velocity of a particle in SHM	$v = - \omega A \sin \omega t$	The velocity (v) of a particle in a transverse wave in SHM as a function of time t. Note: at $t = 0$, $v = 0$.
acceleration of a particle in SHM	$a = - \omega^2 A \cos \omega t$	The acceleration (a) of a particle in a transverse wave in SHM as a function of time t. Note: at $t = 0$, the acceleration is a maximum but the direction is opposite that of the displacement.
velocity of a transverse wave on a string	$v = [F_T/(m/L)]^{1/2}$	The velocity of a transverse wave on a string is related to the tension (F_T) and the mass per unit length (m/L) of the string.
velocity of a longitudinal wave	solid rod $v = (E/\rho)^{1/2}$ liquid or gas $v = (B/\rho)^{1/2}$	The velocity of a longitudinal wave through a solid depends on the elastic modulus (E) and the density of the medium (ρ). For a wave traveling through a liquid or gas, the velocity depends on the **bulk modulus** (B) and the density (ρ).
energy (E) transmitted by a wave	$E = 2 \pi^2 \rho A v t f^2 x_o^2$	The energy (E) transmitted by a wave depends on the: density of the medium (ρ), cross-sectional area (A) through which the wave travels, distance (vt) the wave travels in time t, frequency (f), and amplitude x_o.
average power of the wave	$\bar{P} = 2 \pi^2 \rho A v f^2 x_o^2$	Average power is the average rate of energy transfer.
intensity (I) of a wave	$I = 2 \pi^2 v \rho f^2 x_o^2$	The intensity (I) of a wave is the power transmitted across a unit area (A) perpendicular to the direction of energy.
frequencies of the harmonics of standing waves on a string	$f_n = v/\lambda_n = v/(2L/n)$	The frequency of the particular harmonic depends of the velocity (v) of the wave, the length of the string, and the number of the harmonic (n = 1, 2, 3, etc.).

CONCEPT SUMMARY

Simple Harmonic Motion

Simple harmonic motion (SHM) refers to periodic vibrations or oscillations that exhibit two characteristics: 1) the force acting on the object and the magnitude of the object's acceleration are always directly proportional to the displacement of the object from its equilibrium position, and 2) both the force vector and the acceleration vector are directed opposite to the displacement vector and therefore in toward the object's equilibrium position. The force acting on an object undergoing SHM follows Hooke's law: $\mathbf{F} = -k\,\mathbf{x}$, where **F** is the restoring force acting on the object, k is a proportionality constant called the force constant, and x is the magnitude of the object's displacement from equilibrium. At the equilibrium position the net force on the object equals zero. The negative sign indicates that the force and displacement vectors are oppositely directed.

Amplitude, Frequency, and Period

The **amplitude** (A) of an object undergoing SHM refers to the maximum displacement of the object from the equilibrium position. The **period** (T) of the motion is the time required for the motion to repeat while the **frequency** (f) refers to the number of complete repetitions of the motion that occur each second. The frequency is inversely related to the period, $f = 1/T$.

Energy in the Simple Harmonic Oscillator

The potential energy stored in a vibrating system undergoing SHM is given by $PE = \frac{1}{2} k x^2$ and was previously discussed in Chapter 6 of the textbook. The total energy stored equals the sum of the kinetic energy and potential energy. Assuming that no energy is dissipated due to friction, this total energy remains constant.

$$E = KE + PE = \tfrac{1}{2} m v^2 + \tfrac{1}{2} k x^2$$

At maximum displacement from equilibrium, $v = 0$ and $x = A$. At this point $E = \frac{1}{2} k A^2$. When the displacement equals zero, i.e., $x = 0$, all of the energy is in the form of kinetic energy and $E = \frac{1}{2} m v^2$.

Using the energy principle, it is possible to show that the velocity of the object at any point in its motion can be determined from the equation

$$v = v_o (1 - x^2/A^2)^{1/2} \quad \text{or} \quad v = [(k/m)(A^2 - x^2)]^{1/2}$$

v is the velocity of the object at a displacement x from equilibrium, v_o is the maximum velocity of the object. Maximum velocity occurs as the object is passing through the equilibrium position. A is the amplitude of the motion.

TEXT QUESTION 2. Is the acceleration of a simple harmonic oscillator ever zero? If so, where?

ANSWER: One condition for SHM is a restoring force that is proportional to displacement from equilibrium. For a spring the restoring force follows Hooke's law ($F = -k\,x$). At the point where the displacement equals zero, the force exerted by the spring is zero. Based on Newton's second law $a = F/m$, if $F = 0$ then $a = 0$. As an object in SHM passes through the equilibrium position it has maximum speed but zero acceleration.

The Period of SHM: the Reference Circle

Using the circle as a reference, an analogy can be drawn between the one-dimensional back and forth motion of an object undergoing SHM while attached to a spring with one component of the two-dimensional motion exhibited by an object traveling in a circle. As a result of this analogy, the following equation may be derived for the period of an object of mass m attached to a spring of force constant k and undergoing SHM:

$$T = 2\pi\,(m/k)^{1/2}$$

The larger the mass of the attached object, the greater its inertia and the longer its period. The force constant k is related to the "stiffness" of the spring. The greater the "stiffness" of the spring, the greater the magnitude of the restoring force and acceleration at various displacements and the shorter the period.

EXAMPLE PROBLEM 1. In the diagram shown below, a vertically hung spring stretches 0.300 m when a 1.00 kg object is attached to the end. The object is then displaced 0.120 m from the equilibrium position and released. After release, the object exhibits SHM. Determine the a) force constant of the spring. b) period of the motion, c) total energy of the system, and d) object's speed when it is 0.0700 m from equilibrium.

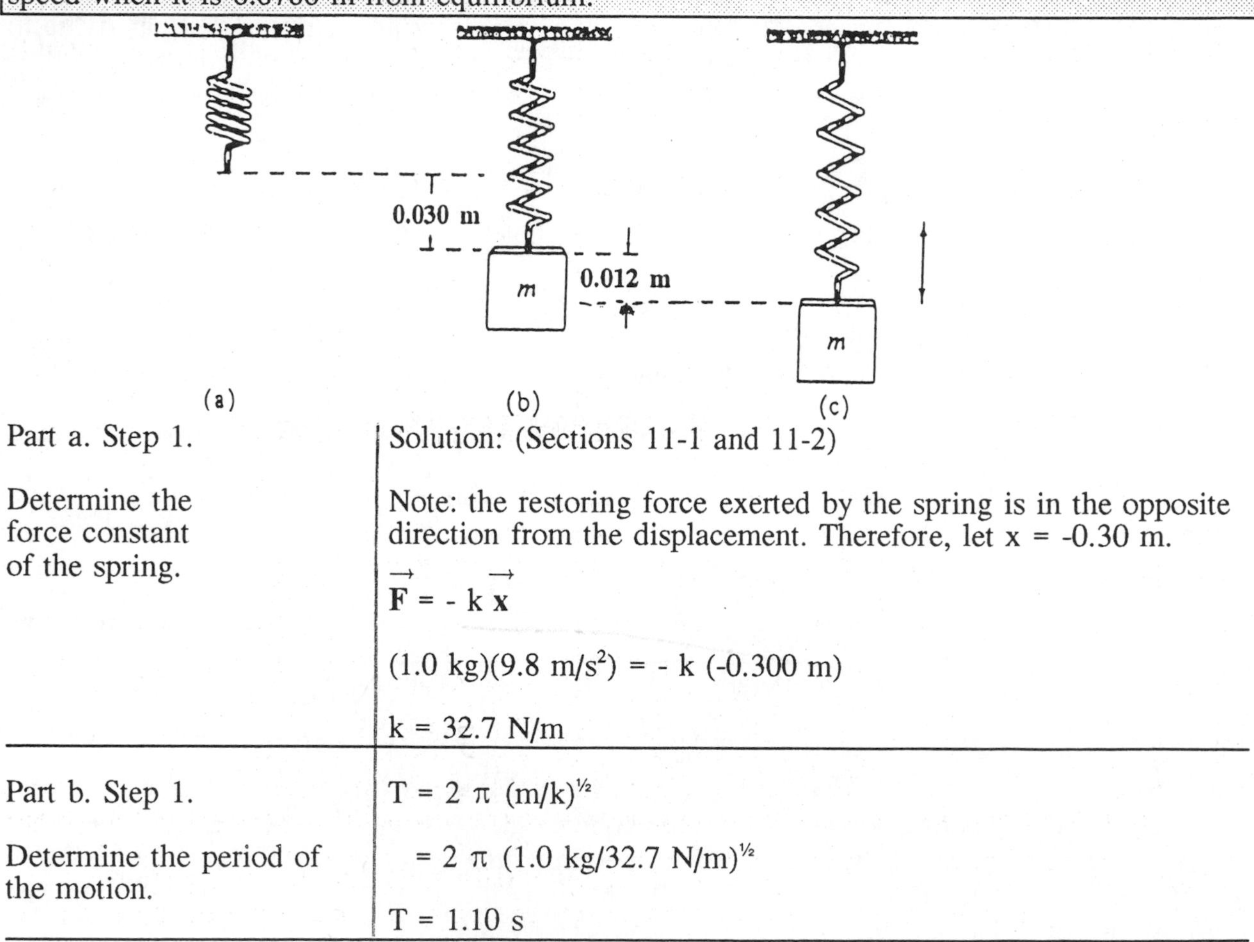

Part a. Step 1. Determine the force constant of the spring.	Solution: (Sections 11-1 and 11-2) Note: the restoring force exerted by the spring is in the opposite direction from the displacement. Therefore, let x = -0.30 m. $\vec{\mathbf{F}} = -k\,\vec{\mathbf{x}}$ $(1.0\text{ kg})(9.8\text{ m/s}^2) = -k\,(-0.300\text{ m})$ $k = 32.7\text{ N/m}$
Part b. Step 1. Determine the period of the motion.	$T = 2\pi\,(m/k)^{1/2}$ $= 2\pi\,(1.0\text{ kg}/32.7\text{ N/m})^{1/2}$ $T = 1.10\text{ s}$

Part c. Step 1. Determine the total energy stored in the system.	The total energy of the system remains constant and equals the sum of the kinetic energy and potential energy at any point. The velocity, and therefore the kinetic energy, at maximum displacement of the object equals zero. The total energy equals the potential energy in the spring at this point. $E = \frac{1}{2} m v^2 + \frac{1}{2} k A^2 = 0 \text{ J} + \frac{1}{2} (32.7 \text{ N/m})(0.120 \text{ m})^2$ $E = 0.235 \text{ J}$
Part d. Step 1. Use the law of conservation of energy to solve for the speed at x = 0.0700 m.	$E = \frac{1}{2} m v^2 + \frac{1}{2} k x^2$ $0.235 \text{ J} = \frac{1}{2} (1.0 \text{ kg})\, v^2 + \frac{1}{2} (32.7 \text{ N/m})(0.0700 \text{ m})^2$ $0.235 \text{ J} = 0.5\, v^2 + 0.080 \text{ J}$ $v = 0.557 \text{ m/s}$

SHM is Sinusoidal

Using the motion of a spring undergoing SHM as a reference, it is possible to show that the graph of the displacement of the object with time has the form of a sine or cosine wave, i.e., the position varies as a function of time. If the displacement of the object equals the amplitude (A) at $t = 0$, then the equation for the displacement, velocity, and acceleration as a function of time can be written as follows:

displacement $\quad x = A \cos \omega t$ where $\omega = 2 \pi f$ or $\omega = 2\pi/T$

velocity $\quad v = - \omega A \sin \omega t$

acceleration $\quad a = - \omega^2 A \cos \omega t$

The following graph represents the displacement vs. time for an object undergoing SHM at the end of a spring.

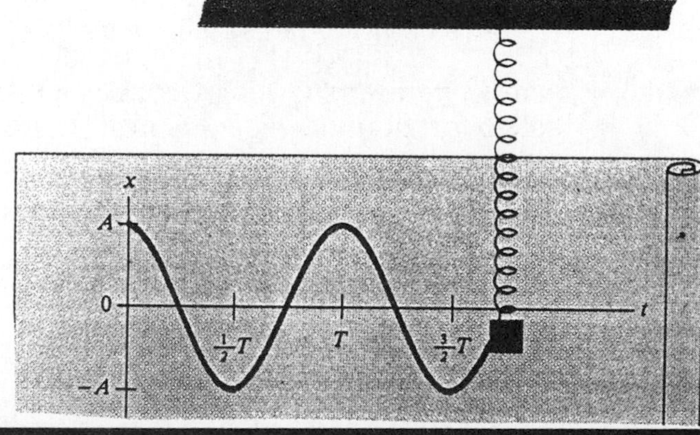

EXAMPLE PROBLEM 2. An object oscillates with SHM according to the equation $x = 2.00 \cos \pi t$ meters, as shown in the above diagram. Determine the a) amplitude, frequency, and period of the motion and b) displacement, velocity, and acceleration of the object at $t = \frac{1}{3}$ s.

Part a. Step 1. Determine the amplitude, SHM, frequency, and period of the motion.	Solution: (Section 11-3) The equation follows the general form of a sinusoidal function in $x = A \cos \omega t$. Thus, the amplitude $A = 2.00$ m. $\omega = 2 \pi f$, then $2 \pi f = \pi$ rad/s $f = (\pi \text{ rad/s})/(2 \pi \text{ rad}) = 0.500$ vibrations/s $f = 0.500$ Hz The period may now be determined: $T = 1/f = 1/(0.500 \text{ vib/s}) = 2.00$ s
Part b. Step 1. Determine the displacement, velocity, and acceleration of the object at $t = ⅓$ second.	Solve for the displacement at $t = ⅓$ s by substituting this value into the displacement equation. $x = A \cos \omega t$ meters $= (2.00 \text{ m}) \cos (\pi \text{ rad/s})(⅓ \text{ s})$ $x = (2.00 \text{ m})(\cos \pi/3 \text{ rad})$ But $(\pi/3 \text{ rad})(360°/2\pi \text{ rad}) = 60.0°$ $x = (2.00 \text{ m}) \cos 60.0° = (2.00 \text{ m})(0.500)$ $x = 1.00$ m Solve for the velocity and acceleration by substituting $t = ⅓$ s into the appropriate equations. $v = - A\omega \sin \omega t$ $= - (2.00 \text{ m})(\pi \text{ rad/s}) [\sin (\pi \text{ rad/s})(⅓ \text{ s})]$ $v = - (2.00 \pi \text{ m/s}) \sin 60.0° = - 5.44$ m/s $a = - A \omega^2 \cos \omega t$ $a = - (2.00 \text{ m/s})(\pi \text{ rad/s})^2 [\cos (\pi \text{ rad/s})(⅓ \text{ s})] = -9.86 \text{ m/s}^2$

Simple Pendulum

A **simple pendulum** is assumed to have its entire mass concentrated at the end of a light string. The simple pendulum undergoes SHM if the maximum angle that it is displaced from equilibrium is small (approximately 15° or less). The formula for the period of motion is

$$T = 2 \pi (L/g)^{½}$$

L is the length of the pendulum and g is the gravitational acceleration. The period of a simple pendulum depends only on its length and the value of g; because of this it can be used as a timing device. The first pendulum clock was built by Christiaan Huygens (1629-1695).

TEXT QUESTION 7. If a pendulum clock is accurate at sea level, will it gain or lose time when taken to high altitude? Why?

ANSWER: The period of a pendulum is given by $T = 2\pi (L/g)^{1/2}$. The period is directly proportional to the length but inversely proportional to the gravitational acceleration. In chapter 5 of the textbook, it was shown that $g = G\,m/r^2$. A pendulum at high altitude is at a greater distance from the center of the earth than at sea level; therefore, g is less than at sea level. If g is less, the period of the pendulum clock is greater and the clock loses time.

EXAMPLE PROBLEM 3. A simple pendulum is used in a physics laboratory experiment to obtain an experimental value for the gravitational acceleration. A particular student measures the length of the pendulum to be 0.600 m, displaces it 10.0° from the equilibrium position, and releases it. Using a stopwatch, the student determines that the period of the pendulum is 1.55 s. Determine the experimental value of the gravitational acceleration.

Part a. Step 1.	Solution: (Section 11-4)
Rearrange the formula for the period of a simple pendulum and and solve for g.	$T = 2\pi (L/g)^{1/2}$ can be used because a 10.0° angle is considered to be a small angle. $T = 2\pi (L/g)^{1/2}$ and $T^2 = 4\pi^2 L/g$, solving for g gives $g = 4\pi^2 L/T^2 = 4\pi^2 (0.600\text{ m})/(1.55\text{ s})^2$ $g = 9.85\text{ m/s}^2$

Damped Harmonic Motion

A system undergoing SHM will exhibit **damping**. Damping is the loss of mechanical energy as the amplitude of motion gradually decreases. In the mechanical systems studied in the previous sections, the losses are generally due to air resistance and internal friction and the energy is transformed into heat.

For the amplitude of the motion to remain constant, it is necessary to add enough mechanical energy each second to offset the energy losses due to damping. A simple example of this is a small child on a swing. If the parent stops giving a periodic push to the swing, its amplitude will gradually decrease until it stops at the equilibrium position.

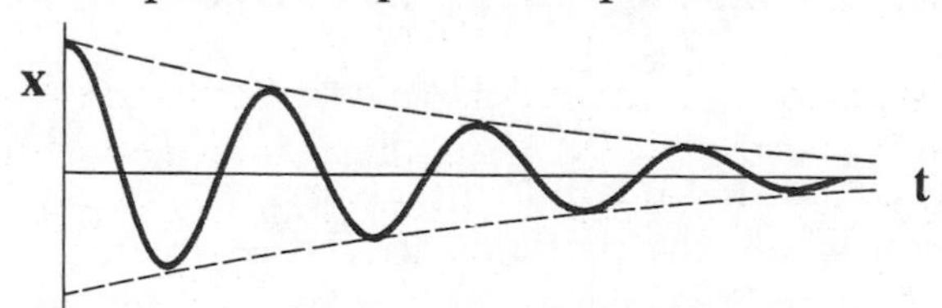

In many instances damping is a desired effect. For example, shock absorbers in a car remove unwanted vibration.

Forced Vibrations: Resonance

An object subjected to an external oscillatory force tends to vibrate. The vibrations that result are called **forced vibrations**. These vibrations have the same frequency as the external force and not the natural frequency of the object.

If the external forced vibrations have the same frequency as the natural frequency of the object, the amplitude of vibration increases and the object exhibits **resonance**. The natural frequency (or frequencies) at which resonance occurs is called the **resonant frequency**. Resonance can be beneficial, e.g., the small periodic additions of energy from the parent to the child on the swing can result in a large amplitude of swing. However, resonance can also be destructive. Proper damping must be included in the design of buildings, bridges, grandstands, etc., to ensure that structural damage does not occur.

Wave Motion

A wave is, in general, a disturbance that moves through a medium. It carries energy from one location to another without transporting the material of the medium. Examples of mechanical waves include water waves, waves on a string, and sound waves.

Transverse Waves

In a **transverse wave**, the particles of the medium move at right angles to the direction of motion of the wave. The top part of the wave is called the **crest** while the portion of the wave below the equilibrium position is called the **trough**. A wave on a rope, as shown below, approximates a transverse wave.

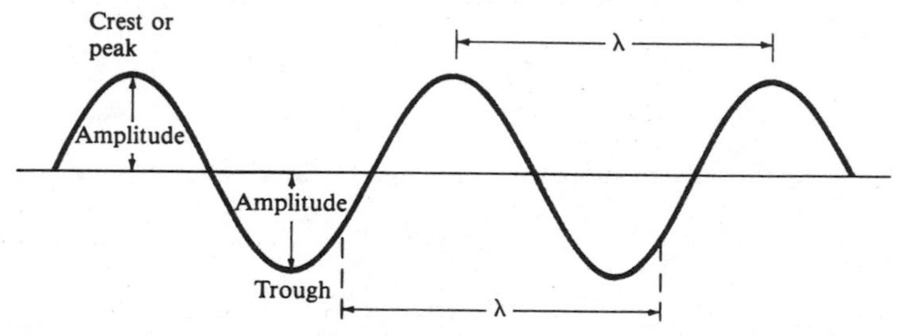

Longitudinal Waves

In a **longitudinal wave,** the particles of the medium move back and forth, parallel to the direction of motion of the wave. The result is a series of expansions alternating with compressions which travel along the length of the medium. The diagram below is a representation of a longitudinal wave traveling along a coiled spring, e.g., a slinky.

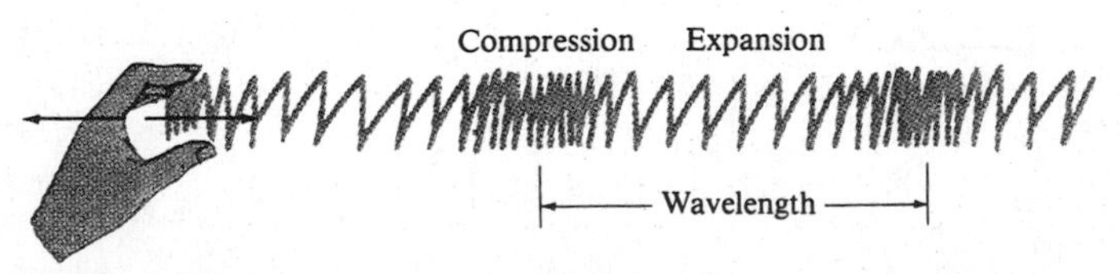

The region where the particles of the medium are close together is called a **compression** (or **condensation**) while the region where the particles are farther apart is called an **expansion** (or **rarefaction**). In analyzing longitudinal waves, it is convenient to compare a region of compression with the crest of a transverse wave and an expansion with the trough of a transverse wave.

Periodic Waves

The velocity (v) of a wave through a medium remains fixed as long as the medium does not change. For a given **periodic** wave, the velocity is given by

$v = f\lambda$

where the frequency (f) is the number of waves passing a particular point each second. The SI unit of frequency is the hertz (Hz) where 1 Hz = 1 wave per second. The wavelength (λ) is the distance between any two repeating points on the wave, e.g., the distance between adjacent crests or adjacent compressions.

The velocity of a wave depends on the properties of the medium through which it travels. For a transverse wave on a string, the velocity is given by the following equation:

$v = [F_T/(m/L)]^{1/2}$

The tension in the string is F_T, while (m/L) is the mass per unit length of the string.

For a longitudinal wave, the velocity depends on the elastic force factor and the density of the medium (ρ). For a longitudinal wave traveling through a solid rod, the elastic force factor is the **elastic modulus** (E). For a longitudinal wave traveling through a liquid or gas, the elastic force factor is the **bulk modulus** (B). The following equations are used to determine the velocity of the wave:

solid rod $v = (E/\rho)^{1/2}$ liquid or gas $v = (B/\rho)^{1/2}$

TEXT QUESTION 13. Explain the difference between the speed of a transverse wave traveling down a cord and the speed of a tiny piece of the cord.

ANSWER: In a transverse wave, the tiny piece moves up and down while the wave moves in a direction perpendicular to the tiny piece. As shown in the diagram on page 11-7, if the wave is a periodic transverse wave, then the tiny piece is moving up and down in SHM. The velocity of the tiny piece is a minimum at the top and bottom of its motion and a maximum as it passes through the equilibrium position.

The wave travels along the cord in a direction perpendicular to the direction of motion of a tiny piece of cord. The speed of the wave along the length of the cord remains constant as long as the tension in the cord and the mass per unit length remain constant.

EXAMPLE PROBLEM 4. Determine a) the speed of sound through a long aluminum rod and b) the wavelength of the sound waves produced by a 440 Hz vibration in the rod. The elastic modulus of aluminum is 70.0×10^9 N/m^2 and the density is 2.70×10^3 kg/m^3.

Part a. Step 1.	Solution: (Section 11-8)
Determine the speed of sound in the aluminum rod.	The speed of sound in a metal rod depends on the elastic modulus (E) and the density (ρ). Substitute the values into the equation for the speed of sound in a metal rod.

	$v = (E/\rho)^{1/2}$ $= [(70.0 \times 10^9 \text{ N/m}^2)/(2.70 \times 10^3 \text{ kg/m}^3)]^{1/2}$ $v = 5090$ m/s
Part b. Step 1. Determine the wavelength of the sound waves in the rod.	The frequency and velocity are now known. Use the equation for the speed of a periodic wave to determine the wavelength. $v = f\lambda$, then 5090 m/s = (440 Hz) λ $\lambda = 11.6$ m

Energy Transmitted by Waves

The energy (E) transmitted by a wave is given by

$$E = 2\pi^2 \rho A v t f^2 x_o^2$$

where A is the cross-sectional area through which the wave travels and not the amplitude of the wave, vt is the distance the wave travels in time t, f is the frequency, and x_o is the amplitude of the wave. The fact that the energy transmitted by the wave is proportional to the square of the frequency and the square of the amplitude is an important result and holds for all types of waves. The average rate of energy transfer is the average power of the wave and is given by

$$\bar{P} = E/t = 2\pi^2 \rho A v f^2 x_o^2$$

The **intensity** (I) of a wave is defined as the power transmitted across a unit area (A) perpendicular to the direction of energy flow and is given by

$$I = \bar{P}/A = 2\pi^2 v \rho f^2 x_o^2$$

As waves travel outward from the center of a disturbance, e.g., an earthquake, the value of the area (A) increases and the intensity and amplitude of the waves decrease. If the wave is in an isotropic medium (same in all directions), the wave is spherical and both intensity and amplitude decrease with the square of the distance from the source. For a sphere, the surface area is $A = 4\pi r^2$.

TEXT QUESTION 18. Two linear waves have the same amplitude and are otherwise identical, except one has half the wavelength of the other. Which transmits more energy? By what factor?

ANSWER: The relation between the energy transported by a wave and the frequency is given by equation $E = 2\pi^2 \rho A v f^2 x_o^2$, where ρ is the density of the medium, A is the cross-sectional area that the wave is passing through, v is the speed of the wave, f is the frequency of the wave, and x_o is the amplitude of the wave.

The frequency (f) of a wave is related to the wavelength by the equation $f = v/\lambda$. Based on this equation the wave that has half the wavelength must have twice the frequency. According to equation $E = 2\pi^2 \rho A v f^2 x_o^2$, the higher frequency wave carries the greater energy. Since the energy is related to the square of the frequency, the shorter wavelength wave has twice the frequency and therefore 2^2 or four times the energy.

Behavior of Waves: Reflection

As shown in the diagram, the wave front of a wave striking a straight barrier follows the **law of reflection**. This law states that the **angle of incidence** equals the **angle of reflection,** $\theta_i = \theta_r$. The angle of incidence is the angle between the direction of the incoming wave and a normal drawn to the surface. The angle of reflection is the angle between the direction of the reflected wave and the normal to the surface.

A transverse wave traveling in a rope undergoes reflection at the point in the rope where the medium changes. If the rope is fixed at one end, as shown in diagram a, a wave crest will undergo a 180° phase change upon reflecting from this point, i.e., a crest will reflect as a trough and vice versa. If the end is free as shown in diagram b, then no phase change occurs and a crest reflects as a crest and a trough reflects as a trough.

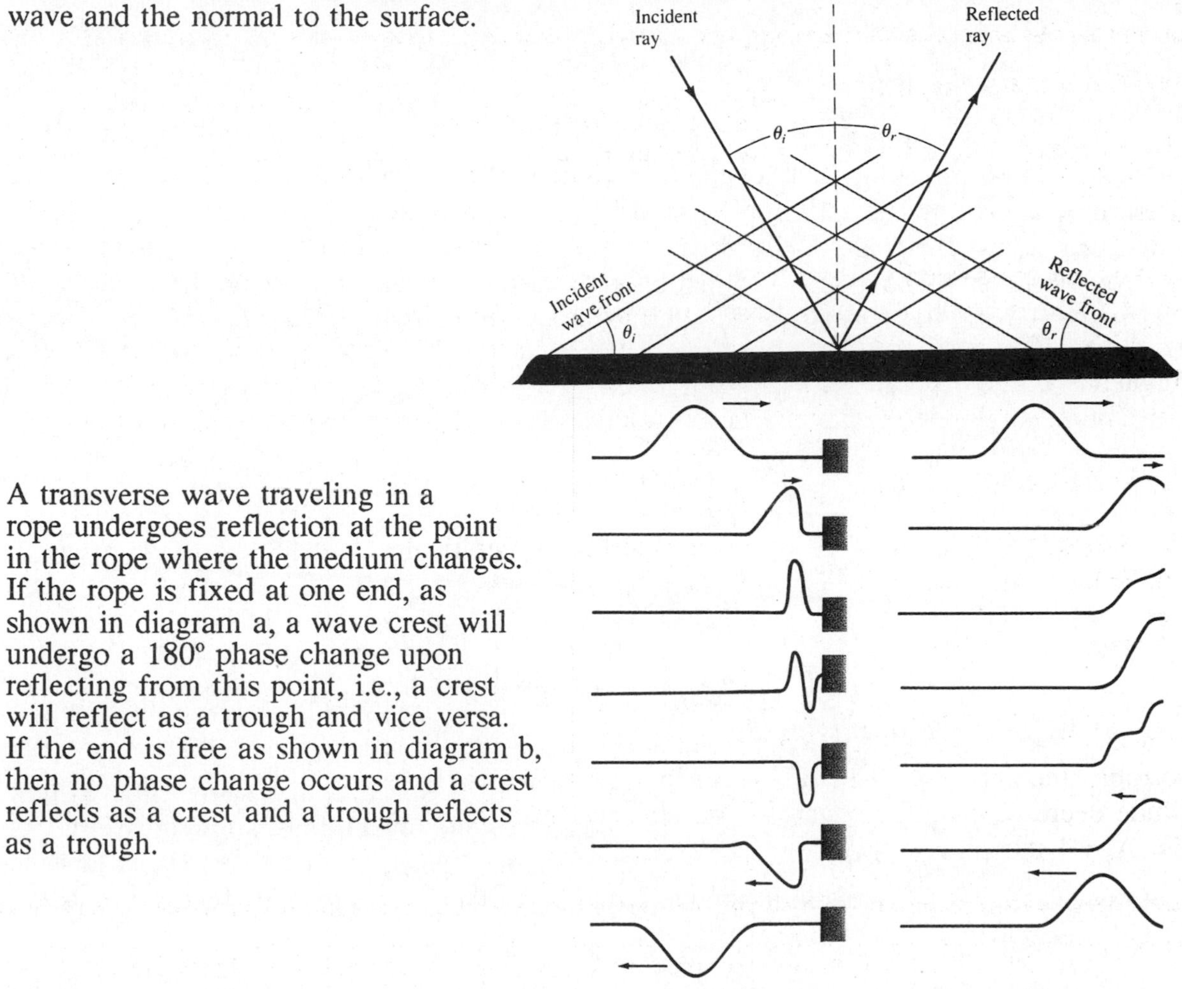

Behavior of Waves: Partial Transmission and Partial Reflection of a Wave Pulse at the Interface of Two Mediums

As shown in the diagram located at the top of the next page, when a crest traveling through a light section of rope reaches a point where it enters a heavier section, then part of the crest is reflected as a trough and part is transmitted as a crest (diagram a). In diagram b, a crest is traveling through a heavy section of rope and reaches a point where it enters a lighter section of rope, then part of the crest is reflected as a crest and part of the crest is transmitted as a crest. A phase change occurs only for that part of the pulse that reflects upon entering a heavier medium.

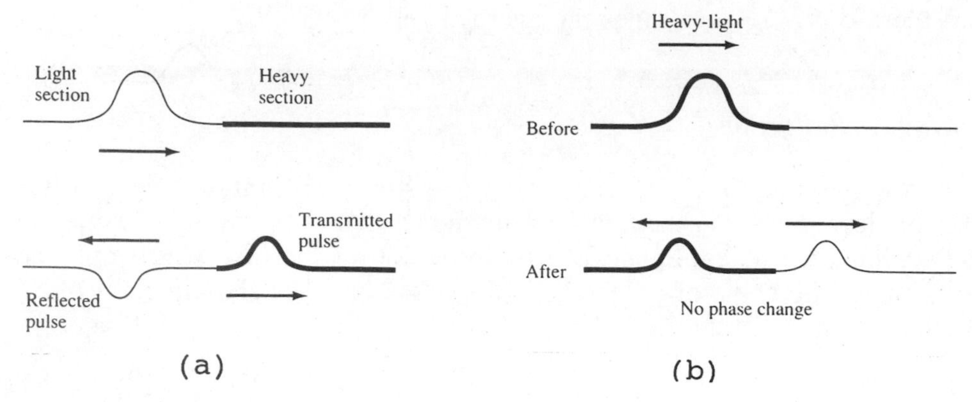

Behavior of Waves: Refraction

A water wave traveling from deep water into shallow water changes direction. This phenomenon is known as **refraction**. The velocity is greater in deep water and, as can be seen in the diagram below, the change in direction is such that the angle of incidence is greater than the angle of refraction. The angle of refraction is defined as the angle between the direction of the refracted wave and the normal to the interface between the two sections.

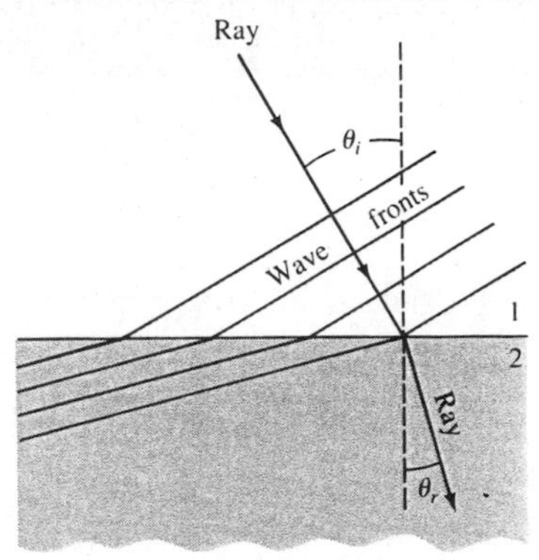

Behavior of Waves: Interference

Two wave pulses passing through the same region of space at the same time exhibit **interference**. The result of this interference is predicted by the **principle of superposition.** As shown in the diagrams a and b located at the top of the next page, this principle states that the resultant displacement of the medium when two waves overlap equals the algebraic sum of their separate displacements.

Destructive interference occurs if the amplitude of the resultant is smaller than either wave pulse. For example, as shown in diagram a, if a crest overlaps with a trough of equal amplitude, e.g., 1 meter, the resulting displacement equals zero, i.e., 1 m - 1 m = 0. After the momentary interaction, the pulses continue moving in their original directions with their original displacements.

Constructive interference occurs if the resultant is greater than the amplitude of either wave pulse. For example, as shown in diagram b, if two wave pulses, each 1 meter high and traveling in opposite directions, interfere, the result will be a momentary displacement of the medium of 2 meters, i.e., 1 m + 1 m = 2 m.

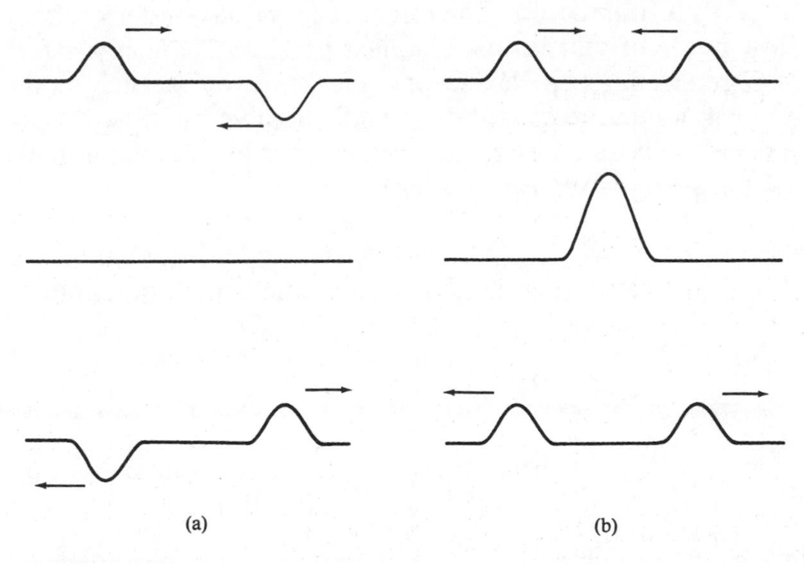

Behavior of Waves: Diffraction

Waves tend to bend as they pass an obstacle. For example, water waves passing a log in a river will tend to bend around the log after they pass. This bending is known as **diffraction.** The amount of diffraction depends on the wavelength of the waves and the size of the obstacle.

Standing Waves in Strings

As shown in the diagram below, a **standing wave** is produced by the superposition of two periodic waves having identical frequencies and amplitudes which are traveling in opposite directions. In stringed musical instruments, the standing wave is produced by waves reflecting off a fixed end and interfering with oncoming waves as they travel back through the medium.

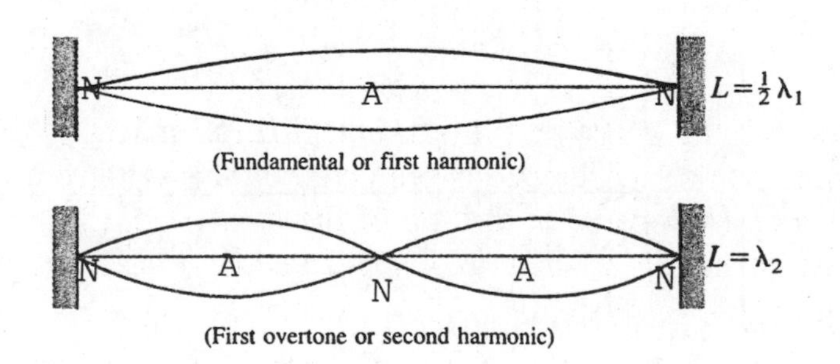

If a string is of fixed length and tension, the standing wave is produced only for certain wavelengths and resonant frequencies. A standing wave has certain points along its entire length where the displacement is always zero. These points are called **nodal points.** The nodal points are found at half wavelength ($\frac{1}{2}\lambda$) intervals along the length of the string.

Points of maximum amplitude, called **antinodal points,** are located halfway between adjacent nodal points. The displacement of the medium at the point fluctuates between a crest and a trough. The amplitude of the wave at the antinodal points equals the sum of the amplitudes of the two component waves that produce the standing wave pattern.

The lowest possible frequency that can produce a standing wave in a string fixed at both ends is called the **first harmonic**. This frequency is also referred to as the **fundamental frequency** or **first mode of vibration**. The next possible frequency is the first overtone. This frequency is also referred to as the second harmonic or second mode of vibration. The frequencies of the harmonics are consecutive whole number multiples of the first harmonic. The positions of the nodal points are designated by the letter N and the antinodal points by the letter A. The length of the string is represented by L.

In general, $\frac{1}{2}\lambda_n = L/n$, where n = 1, 2, 3, 4, etc., and n is the number of vibrating segments in the string. The frequency of the particular harmonic can be determined as follows:

$f = v/\lambda_n = v/(2L/n)$

EXAMPLE PROBLEM 5. When the tension is adjusted to 20.0 N, a 60.0 Hz vibration will produce a third harmonic standing wave in a string 0.250 m long. Determine the a) velocity of the waves in the string and b) mass of the string.

Part a. Step 1. Draw a diagram representing the third harmonic.	Solution: (Section 11-12) N A N A N A N $L = \frac{3}{2}\lambda_3$ (Second overtone or third harmonic)
Part a. Step 2. Use the diagram to calculate the wavelength.	The nodal points are $\frac{1}{2}\lambda$ apart; the wavelength of the waves can be determined. $\frac{1}{2}\lambda_3 = \frac{1}{3}L$ $\lambda_3 = \frac{2}{3}L = \frac{2}{3}(25.0\ m) = 0.167\ m$
Part a. Step 3. Determine the velocity of the periodic waves.	$v = f\lambda_3$ $= (60.0\ Hz)(0.167\ m)$ $v = 10.0\ m/s$
Part b. Step 1. Rearrange the equation that relates the tension in a string to the velocity of the waves and solve for the mass.	$v = [F_T/(m/L)]^{1/2}$ and $v^2 = F_T/(m/L)$ $F_T = v^2\ m/L$ and rearranging gives $m = (F_T\ L)/v^2$ $m = (20.0\ N)(0.250\ m)/(10.0\ m/s)^2 = 5.00 \times 10^{-2}\ kg$

PROBLEM SOLVING SKILLS

For problems involving a spring undergoing SHM:

1. Complete a data table listing the magnitude of the restoring force at maximum displacement, mass of object, amplitude of motion, etc.
2. Use Hooke's law to determine the force constant of the spring.
3. To determine the period, use the formula that relates the period to the force constant and the mass of the object.
4. Determine the total energy of the system and use the law of conservation of energy to determine the velocity of the object at any point in the motion.

For problems involving a simple pendulum undergoing SHM:

1. Use $T = 2\pi (L/g)^{1/2}$ to solve for the period, length, or the magnitude of the gravitational acceleration.

For problems involving the speed of a longitudinal waves passing through a solid, liquid, or gas:

1. If the medium is a long solid rod, note the density and the elastic modulus of the rod and solve for the velocity.
2. If the medium is a liquid or a gas, note the bulk modulus and the density and solve for the velocity.

For problems involving energy transported by a wave, the power of a wave, and wave intensity:

1. Complete a data table and list the velocity, frequency, and amplitude of the wave and the cross-sectional area through which the wave travels.
2. Apply the formula for energy, power, and intensity to solve the problem.

For problems involving standing waves in strings:

1. Complete a data table listing the tension, mass and length of string, and the frequency of the vibration causing the standing wave.
2. Draw an accurate diagram for the harmonic described in the problem. Locate the position of the nodes and antinodes. Determine the wavelength of the wave.
3. If the frequency is known, use $v = f\lambda$ to determine the velocity. If the tension and mass per unit length are known, use $v = [F_T/(m/L)]^{1/2}$ to solve for the velocity.
4. Solve for the unknown quantity.

SOLUTIONS TO SELECTED TEXTBOOK PROBLEMS

TEXTBOOK PROBLEM 5. An elastic cord vibrates with a frequency of 3.0 Hz when a mass of 0.60 kg is hung from it. What will its frequency be if only 0.38 kg hangs from it?

Part a. Step 1.	Solution: (Sections 11-1 to 11-3)
Determine the period of the vibration.	$T = 1/f = 1/(3.0\ Hz) = 0.33\ s$

Part a. Step 2. Determine the force constant of the spring.	$T = 2\pi (m/k)^{1/2}$ and rearranging gives $k = 4\pi^2 m/T^2$ $k = 4\pi^2 (0.60 \text{ kg})/(0.33 \text{ s})^2$ $k = 220$ N/m
Part a. Step 3. Determine the frequency if the mass is 0.38 kg.	$T = 2\pi (m/k)^{1/2}$ $= 2\pi [(0.38 \text{ kg})/(220 \text{ N/m})]^{1/2}$ $T = 0.26$ s
Part a. Step 4. Determine the frequency of the vibration.	$T = 1/f$ and $f = 1/T$ $f = 1/(0.26 \text{ s}) = 3.8$ Hz

TEXTBOOK PROBLEM 23. At t = 0, a 755 g mass at rest on the end of a horizontal spring (k = 124 N/m) is struck by a hammer, which gives the mass an initial speed of 2.96 m/s. Determine (a) the period and frequency of the motion, (b) the amplitude, (c) the maximum acceleration, (d) the position as a function of time, and (e) the total energy.

Part a. Step 1. Determine the period and frequency the motion.	Solution: (Sections 11-1 to 11-3) $T = 2\pi (m/k)^{1/2}$ $= 2\pi (0.755 \text{ kg}/124 \text{ N/m})^{1/2}$ $T = 0.490$ s $f = 1/T = 1/(0.490 \text{ s}) = 2.04$ Hz
Part b. Step 1. Determine the amplitude of the motion.	The velocity is a maximum when the object is first struck. The point at which it is struck is the equilibrium position ($x_1 = 0$ m) of the resulting simple harmonic motion. At $x_1 = 0$ m, $v = 2.96$ m/s. At maximum displacement $x_2 = A$ and $v_2 = 0$. $KE_1 + PE_1 = KE_2 + PE_2$ $\frac{1}{2} m v_1^2 + \frac{1}{2} k x_1^2 = \frac{1}{2} m v_2^2 + \frac{1}{2} k x_2^2$ Simplifying gives $m v_1^2 = k A^2$ $A^2 = (m/k) v_1^2 = [(0.755 \text{ kg})/(124 \text{ N/m})](2.96 \text{ m/s})^2 = 0.0533 \text{ m}^2$ $A = 0.231$ m

Part c. Step 1. Determine the maximum acceleration.	$F = - k x$ and $F = m a$ The acceleration is a maximum when x = A. $m a = - k x$ $a = - (k/m) A = - [(124 N/m)/(0.755 kg)](0.231 m)$ $a = -37.9 m/s^2$ Note: the negative sign indicates that the acceleration and displacement vector are in opposite directions.
Part d. Step 1. Determine the position as a function of time.	At t = 0, x = 0. Therefore, the motion follows a sine function. $x = A \sin \omega t$ where $\omega = 2 \pi f = 2 \pi (2.04 Hz) = 12.8 rad/s$ $x = (0.231 m) \sin (12.8 rad/s) t$
Part e. Step 1. Determine the total energy stored in the system.	The total energy of the system remains constant and equals the sum of the kinetic energy and potential energy at any point. The velocity was a maximum when the potential energy was zero. Therefore, the total energy stored in the system equals the initial kinetic energy. $E_{total} = ½ m v_1^2 + ½ k x_1^2$ when $x_1 = 0 m$, $v_1 = 2.96 m/s$ $E_{total} = ½ (0.755 kg)(2.96 m/s)^2 + ½ (124 N/m)(0 m)^2$ $E_{total} = 3.31 J$

TEXTBOOK PROBLEM 41. A cord of mass 0.65 kg is stretched between two supports 30 m apart. If the tension in the cord is 150 N, how long will it take a pulse to travel from on support to the other?

Part a. Step 1. Determine the mass per unit length of the string.	Solution: (Sections 11-7 and 11-8) $m/L = (0.65 kg)/(30 m) = 2.17 \times 10^{-2} kg/m$
Part a. Step 2. Determine the velocity of the pulse as it travels through the string.	$v = [F_T/(m/L)]^{½}$ $v = [(150 N)/(2.17 \times 10^{-2} kg/m)]^{½} = 83.2 m/s$
Part a. Step 3. Determine the time for the pulse to travel from one support to the other.	time = distance/speed = (30 m)/(83.2 m/s) time = 0.36 s

TEXTBOOK PROBLEM 47. The intensity of an earthquake wave is measured to be 2.0 x 10^6 $J/m^2 \cdot s$ at a distance of 48 km from the source. (a) What was the intensity when it passed a point only 1.0 km from the source? (b) At what rate did energy pass through an area of 5.0 m^2 at 1.0 km?

Part a. Step 1.

Determine the intensity at a point 1.0 km from the source.

Solution: (Section 11-9)

Assume that the energy of the earthquake spreads out uniformly in all directions. The intensity is related to the power by $I = P/A$, where the area is given by $A = 4 \pi r^2$ and r is the distance from the source. Therefore, the intensity is diminished by the square of the distance from the origin.

$$I_{1.0\ km}/I_{50\ km} = [4 \pi (48\ km)^2]/[4 \pi (1.0\ km)^2]$$

$$I_{1.0\ km} = (2300)(2.0 \times 10^6\ J/m^2 \cdot s) = 4.6 \times 10^9\ J/m^2 \cdot s$$

Part b. Step 1.

Determine the rate energy passed through an area of 5.0 m^2 at 1.0 km.

The rate at which energy passes through an area equals the power.

$$P = I A = (4.6 \times 10^9\ J/m^2 \cdot s)(5.0\ m^2)$$

$$P = 2.3 \times 10^{10}\ J/s = 2.3 \times 10^{10}\ W$$

TEXTBOOK PROBLEM 57. A guitar string is 90 cm long and has a mass of 3.6 g. The distance from the bridge to the support post is L = 62 cm, and the string is under a tension of 520 N. What are the frequencies of the fundamental and first two overtones?

Part a. Step 1

Draw diagrams of the fundamental and first two overtones.

Solution: (Section 11-13)

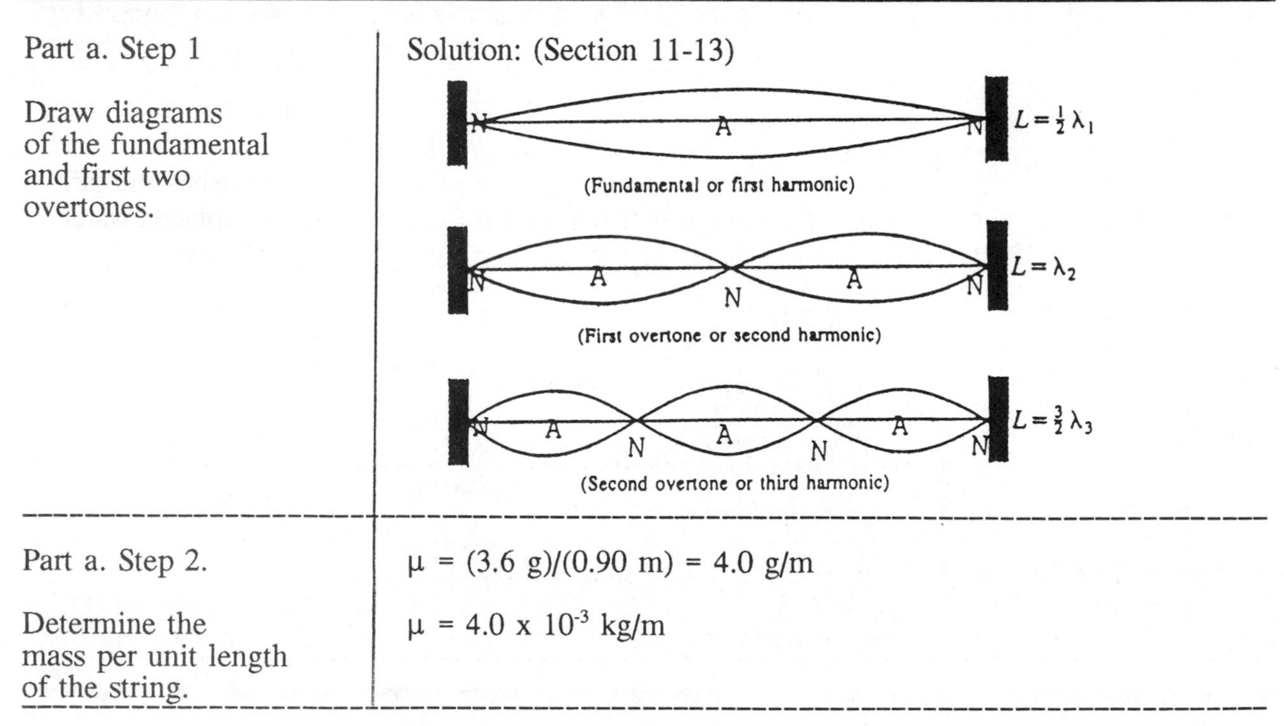

(Fundamental or first harmonic)

(First overtone or second harmonic)

(Second overtone or third harmonic)

Part a. Step 2.

Determine the mass per unit length of the string.

$$\mu = (3.6\ g)/(0.90\ m) = 4.0\ g/m$$

$$\mu = 4.0 \times 10^{-3}\ kg/m$$

Part a. Step 3. Determine the speed of the waves as they travel through the string.	$v = (F_T/\mu)^{1/2}$ $= [(520\ N)/(4.0 \times 10^{-3})]^{1/2}$ $v = 361$ m/s
Part a. Step 4. Determine the wave-length of the waves needed to produce the fundamental frequency.	The distance between adjacent nodal points in a standing wave equals one-half the wavelength of the wave. For the fundamental frequency $\frac{1}{2} \lambda = 0.62$ m and $\lambda = 1.24$ m
Part a. Step 5. Determine the fundamental frequency.	$v = f \lambda$ 361 m/s $= f_1$ (1.24 m) $f_1 = 290$ Hz
Part a. Step 6. Determine the frequencies of the first two overtones.	The frequencies of the overtones are consecutive integer multiples of the fundamental, i. e. $f_n = n f_1$ where n = 1, 2, etc. Therefore, $f_2 = 2(290\ Hz) = 580$ Hz $f_3 = 3(290\ Hz) = 870$ Hz

TEXTBOOK PROBLEM 76. A 220 kg wooden raft floats on a lake. When a 75 kg man stands on the raft, it sinks 4.0 cm deeper into the water. When he steps off, the raft vibrates for a while. (a) What is the frequency of vibration? (b) What is the total energy of vibration (ignoring damping)?

Part a. Step 1. Determine the force constant.	Solution: (Sections 11-1 and 11-2) When the 75 kg man steps on the raft, his weight pushes the raft further into the water. The buoyant force due to the displaced water acts like the restoring force of a spring on the man + raft. $\mathbf{F} = - k \Delta y$ $(75\ kg)(9.8\ m/s^2) = - k\ (-0.040\ m)$ $k = 1.84 \times 10^4$ N/m
Part a. Step 2. Determine the period of motion of the 220 kg raft.	$T = 2 \pi (m/k)^{1/2}$ $= 2 \pi [(220\ kg)/(1.84 \times 10^4\ N/m)]^{1/2}$ $T = 2 \pi (0.11\ s) = 0.69$ s

Part a. Step 3. Determine the frequency of the vibration.	$f = 1/T = 1/(0.69\ s)$ $f = 1.4\ Hz$
Part b. Step 1. Determine the total energy of vibration.	The total energy of vibration equals the spring potential energy stored when the man is standing on the raft. $PE = \frac{1}{2} k (\Delta y)^2$ $PE = \frac{1}{2} (1.84 \times 10^4\ N/m)(0.040\ m)^2 = 15\ J$

CHAPTER 12

SOUND

OBJECTIVES

After studying the material of this chapter, the student should be able to:

- determine the speed of sound in air at one atmosphere of pressure at different temperatures.
- distinguish between the following terms: pitch, frequency, wavelength, sound intensity, and loudness.
- determine intensity level in decibels of a sound if the intensity of the sound is given in W/m^2.
- explain how a standing wave can be produced in a wind instrument open at both ends or closed at one end and calculate the frequencies produced by different harmonics of pipes of given length.
- determine the beat frequency produced by two tuning forks of different frequencies.
- explain how an interference pattern can be produced by two sources of sound of the same wavelength separated by a distance d.
- solve problems involving two sources for m, d, λ, and the angular separation (θ) when the other quantities are given.
- solve for the frequency of the sound heard by a listener and the wavelength of the sound between a source and the listener when the frequency of the sound produced by the source and the velocity of both the source and the listener are given.
- explain how a shock wave can be produced and what is meant by the term sonic boom.

KEY TERMS AND PHRASES

sound is a longitudinal wave produced by a vibration which travels away from the source through solids, liquids, or gases, but not through a vacuum. Since a sound wave is a longitudinal wave there are regions of compression (condensation) and expansion (rarefaction) as the wave moves through the medium that transports it.

pitch refers to the frequency of a sound wave and is measured in hertz (Hz).

intensity level of sound is the energy transported by a wave per unit time per unit area. Intensity level of sound is measured in bels; however, the decibel (dB) is more commonly used.

open pipe is a wind instrument that is open to the air at both ends. The first harmonic standing wave that produces resonance has an antinode near each end and one node in the middle of the air column.

closed pipe is a wind instrument that is open to the air at one end but closed at the other end. The first harmonic standing wave that produces resonance in the pipe has an antinode near the open end and a node at the closed end.

sound quality depends on the presence of overtones, their number, and relative amplitudes. The result gives each musical instrument its characteristic quality and timbre.

beats are regular pulsations in the loudness of a sound due to two waves of equal amplitude but slightly different frequencies traveling in the same direction. The number of beats per second is equal to the difference between the frequencies of the component waves and is known as the **beat frequency.**

Doppler effect refers to the perceived change of frequency of a wave when there is relative motion between the source and the listener.

shock waves and **sonic booms** are produced when the speed of the source of sound exceeds the speed of sound. The sound waves in front of the source tend to overlap and constructively interfere. The superposition of the waves produce an extremely large amplitude, high energy wave called a shock wave. When the shock wave passes a listener, this energy is heard as a sonic boom.

SUMMARY OF MATHEMATICAL FORMULAS

speed of sound in air measured in meters per second	$v = 331 + 0.6\ T$ m/s	The speed of sound (v) in air changes with temperature. 331 m/s is the speed of sound in air at 0°C and T is the temperature of air in degrees centigrade.
sound intensity level	$\beta(\text{in dB}) = 10 \log I/I_o$	β is the intensity level in decibels. I is the intensity of the sound in W/m^2. The threshold of hearing $I_o = 1.0 \times 10^{-12}\ W/m^2$.
harmonic frequency produced by an open pipe	$f_n = v/\lambda_n = v/(2L/n)$	The harmonic frequency produced by an open pipe depends on the speed of sound (v) and the length of the pipe (L). The frequencies of successive harmonics are consecutive whole number multiples of the first harmonic, i.e., n = 1, 2, 3, etc.
harmonic frequency produced by a pipe closed at one end	$f_n = v/\lambda_n$ $f_n = v/[4L/(2n - 1)]$	The harmonic frequency produced by a closed pipe depends on the speed of sound (v) and the length of the pipe (L). The frequencies of successive harmonics are consecutive odd whole number multiples of the first harmonic, i.e., n = 1, 3, 5, 7, etc.

sound interference pattern produced by two sources of sound which are in phase and have the same wavelength	constructive interference $\sin\theta = m\lambda/d$ destructive interference $\sin\theta = (m + ½)\lambda/d$	angular displacement (θ) of maxima and minima produced by two sources of sound which are in phase and have the same wavelength (λ). m = 1, 2, 3, etc.
beat frequency	$f_b = f_1 - f_2$	The beat frequency equals the difference between the two frequencies.
Doppler effect source moving toward a stationary listener source moving away from a stationary listener listener moving toward a stationary source listener moving toward a stationary source	$f' = [1/(1 - v_s/v)]\,f$ $f' = [1/(1 + v_s/v)]\,f$ $f' = [1 + (v_o)/(v)]\,f$ $f' = [1 - (v_o)/(v)]\,f$	The frequency (f') heard by the listener depends on the speed of sound (v), the frequency of the sound emitted by the source (f), speed of the source (v_s) or the speed of the listener (v_o), and the direction of motion of the source (or listener).

CONCEPT SUMMARY

Sound Waves

Sound is a longitudinal wave produced by a vibration which travels away from the source through solids, liquids, or gases, but not through a vacuum. Since a sound wave is a longitudinal wave there are regions of compression (condensation) and expansion (rarefaction) as the wave moves through the medium that transports it.

The speed is independent of the barometric pressure, frequency, and wavelength of the sound. However, the speed of sound in a gas is proportional to the temperature. Use the following equation to determine the speed of sound in air:

$v = 331 + 0.6\,T$ m/s

where v is the speed of sound in meters per second, 331 is the speed of sound in m/s at 0°C, and T is the temperature in degrees centigrade.

Pitch

Pitch refers to the frequency of a sound wave and is measured in hertz (Hz). The range of human hearing is from 20 Hz to 20,000 Hz. This is called the audible range. This range is considered to be the limits of human hearing. The actual range that can be heard varies from person to person. As a person ages, the ear becomes less responsive to higher frequencies.

Sound waves above 20,000 Hz are called ultrasonic and have found application in medicine and other fields. Those below 20 Hz are called infrasonic waves.

Sound Intensity

The ear transforms the energy of sound waves into electrical signals which are carried to the brain by the nerves. The ear is not equally responsive to all frequencies of sound. It is most sensitive to sounds between 2000 Hz and 3000 Hz.

Loudness is a subjective physiological sensation in a human being that increases with the intensity of a sound. Our subjective sensation of loudness depends not only on **intensity** but also on frequency. A person easily hearing a sound at 1000 Hz may not be able to hear a sound of equal intensity at 50 Hz.

The loudness of a sound is approximately proportional to the common logarithm of the intensity. The intensity level is measured in **bels**, named after Alexander Graham Bell; however, the **decibel** (dB) is more commonly used. The decibel equals one-tenth bel, i.e., 1 dB = 0.1 bel. The formula for the intensity level is

$$\beta(\text{in dB}) = 10 \log I/I_o$$

where β is the intensity level in decibels, I is the intensity of the sound in watt/m^2, and I_o is the minimum intensity audible to the average person and is called the threshold of hearing.

$$I_o = 1.0 \times 10^{-12} \text{ watt/m}^2.$$

The sound levels in dB that are common in everyday life extend from 0 dB (the threshold of hearing) to 140 dB for a jet plane 30 meters away. Ordinary conversation is approximately 65 dB while an indoor rock concert may be 120 dB, which is at the threshold of pain. Exposure to sound levels above 85 dB over an extended period of time may lead to permanent damage to a person's hearing. A table listing intensity levels and intensities of various sounds is given in the textbook.

TEXTBOOK QUESTION 2. What is the evidence that sound is a form of energy?

ANSWER: A source of a sound forces the molecules of the surrounding medium to move in vibratory motion. From chapter 6 in the textbook, a force acting on an object which moves the object through a distance does work. This work equals the change in the object's energy. Evidence of sound energy is apparent to any student who has ever attended a rock concert. If the music is loud enough, the student can feel his chest moving to the beat of the music.

EXAMPLE PROBLEM 1. During an indoor rock concert, sound intensities of 0.15 W/m^2 are produced. Determine the a) intensity levels of these sounds in decibels, and b) amplitude of the sound wave if its frequency is 500 hertz. The density of air is 1.29 kg/m^3 and the speed of sound is 340 m/s.

Part a. Step 1. Apply the formula for the intensity level in decibels.	Solution: (Section 12-2) $I_o = 1.0 \times 10^{-12}$ W/m^2. $\beta = 10 \log (I/I_o)$ $= 10 \log (0.15\ W/m^2/1.0 \times 10^{-12}\ W/m^2) = 10 \log (1.5 \times 10^{11})$ Using your calculator, $\log (1.5 \times 10^{11}) = 11.2$ therefore, $\beta = (10)(11.2)\ dB = 112\ dB$
Part b. Step 1. Use the equation introduced in chapter 11 of the textbook that relates intensity, frequency, and amplitude to solve for the amplitude.	$I = 2\ \pi^2\ v\ \rho\ f^2\ x_o^2$ $0.15\ W/m^2 = 2\ \pi^2 (340\ m/s)(1.29\ kg/m^3)(500\ Hz)^2\ x_o^2$ $x_o^2 = 6.94 \times 10^{-11}\ m^2$ $x_o = 8.3 \times 10^{-6}$ m or 0.0083 mm Even for very loud sounds, the maximum displacement of air from the equilibrium position is very small.

Sources of Sounds: Musical Instruments

String Instruments

While the theory behind standing waves in stringed instruments was discussed in Chapter 11 of the textbook, it should be noted that the sounds produced by vibrating strings are not very loud. Thus, many stringed instruments make use of a sounding board or box, sometimes called a resonator, to amplify the sounds produced. The strings on a piano are attached to a sounding board while for guitar strings a sound box is used. When the string is plucked and begins to vibrate, the sounding board or box begins to vibrate as well. Since the board or box has a greater area in contact with the air, it tends to amplify the sounds.

Wind Instruments

The sounds that are produced by a **wind instrument** are the result of standing waves produced in the air contained within the instrument. In some wind instruments, such as woodwinds or brasses, the air is set into vibration by a vibrating reed or the vibrating lip of the musician. In other cases, e.g., the flute or organ, a stream of air is directed against one edge of an opening or mouthpiece. The resulting turbulence produces vibrations within the instrument. The vibrations cover a range of frequencies which are due to longitudinal standing waves which are created in the air column.

Open Pipe

A wind instrument that is open to the air at both ends is known as an **open tube** or **pipe**. The longitudinal standing wave that produces the sound has an antinode at each end and at least one node in the air column. Assuming that the speed of sound is constant, and $v = f\ \lambda$ for a periodic wave, the frequency produced depends on the length of the tube.

The possible modes of vibration, called harmonics, are similar to those produced in strings. In order to simplify the discussion, the diagrams below show transverse standing waves rather than longitudinal standing waves. The antinodal point in a longitudinal standing wave would actually be a region of alternating compressions and expansions and high pressure variation while at the nodal points the air pressure would remain relatively constant.

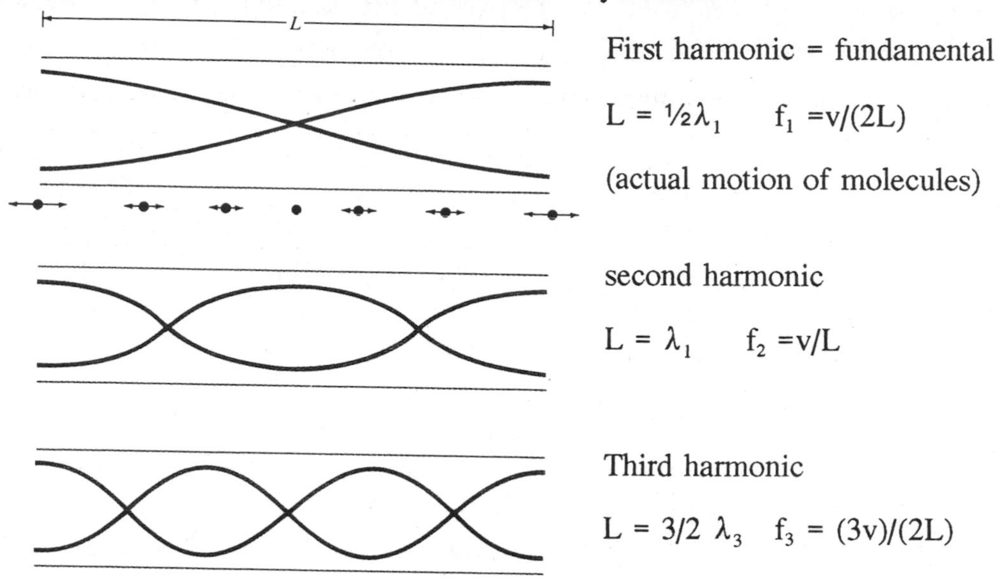

In general, $½\lambda_n = L/n$, where n is an integer and refers to the mode of vibration, e.g., for the 3rd harmonic $n = 3$. The frequency of the particular harmonic can then be determined since $f = v/\lambda_n = v/(2L/n)$. The frequencies of successive harmonics are consecutive whole number multiples of the first harmonic, i.e., 2, 3, 4, etc., times the frequency of the first harmonic.

Closed Pipe

A wind instrument that is open to the air at one end but closed at the other end is known as a **closed tube** or **closed pipe**. The longitudinal standing wave produced in the pipe has an antinode at the open end but a node at the closed end. Assuming the speed of sound is constant, and $v = f\ \lambda$ for a periodic wave, then the frequency produced depends on the length of the tube.

The diagram shown below represents the first mode of vibration in a closed pipe. The second and third modes of vibration are shown at the top of the next page.

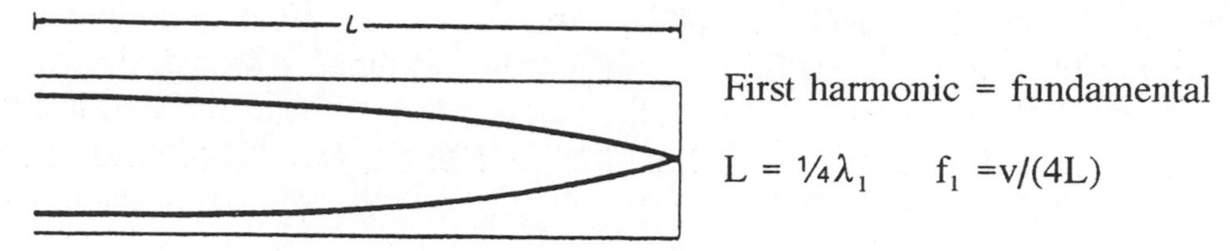

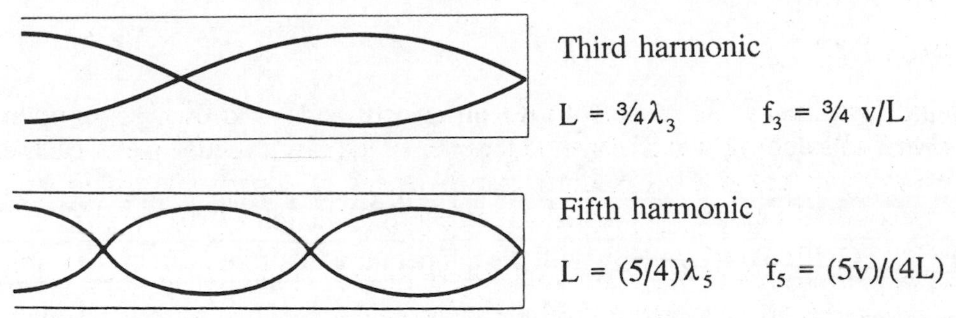

In general, $\frac{1}{4}\lambda_n = L/(2n - 1)$, where n is an integer and refers to the mode of vibration, e.g., for the third harmonic, n = 3. The frequency of the particular harmonic can then be determined, since $f_n = v/\lambda_n = v/[4L/(2n - 1)]$. The frequencies of successive harmonics for a pipe closed at one end are consecutive odd number multiples of the first harmonic, i.e., 3, 5, 7, etc., times the frequency of the first harmonic.

EXAMPLE PROBLEM 2. (a) An open pipe produces a third harmonic standing wave of frequency 1000 Hz on a day when the speed of sound is 340 m/s. Determine the length of the pipe. (b) The pipe is now closed at one end, determine the frequency of the third harmonic.

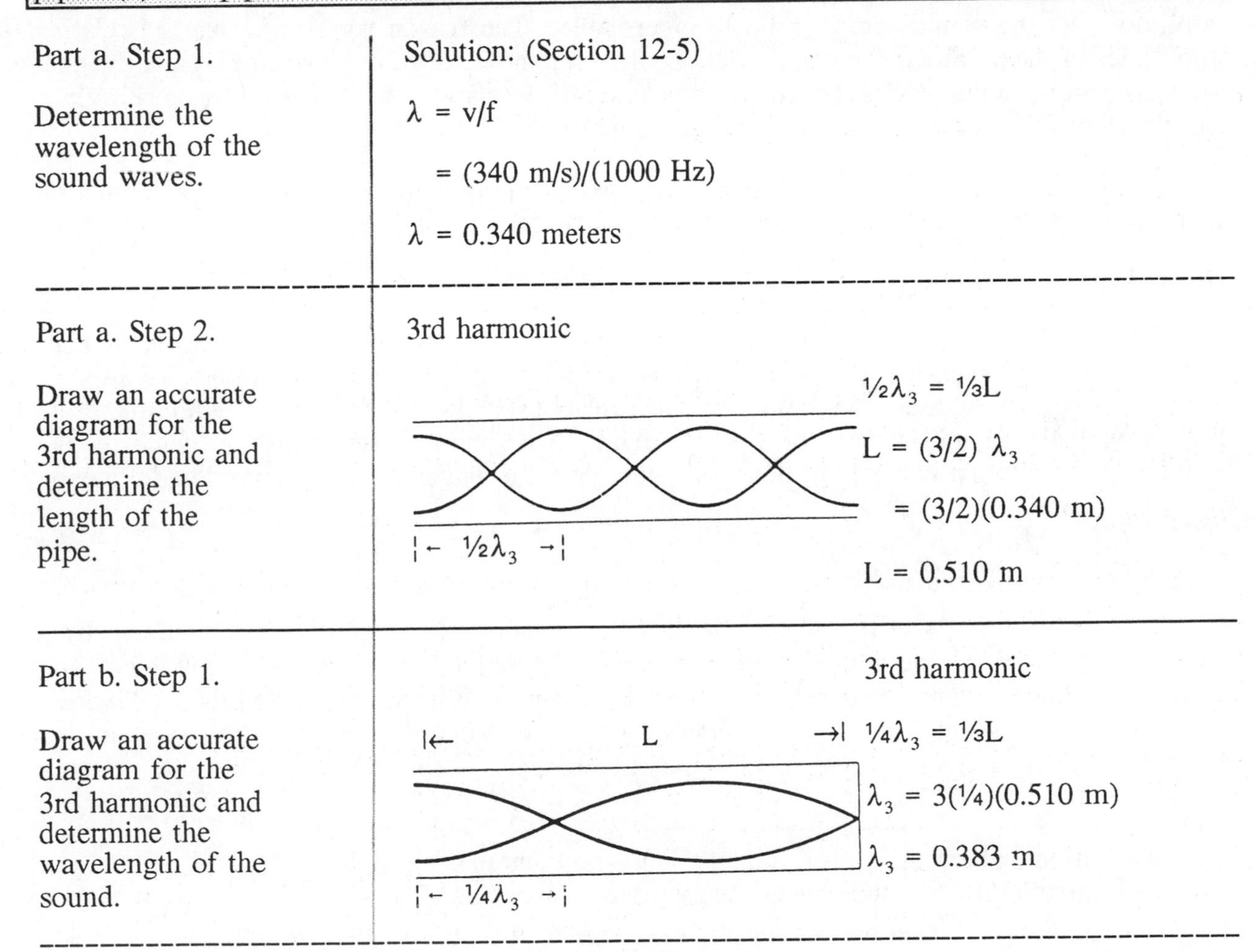

	Solution: (Section 12-5)
Part a. Step 1. Determine the wavelength of the sound waves.	$\lambda = v/f$ $= (340\text{ m/s})/(1000\text{ Hz})$ $\lambda = 0.340$ meters
Part a. Step 2. Draw an accurate diagram for the 3rd harmonic and determine the length of the pipe.	3rd harmonic $\vert\leftarrow \frac{1}{2}\lambda_3 \rightarrow\vert$ $\frac{1}{2}\lambda_3 = \frac{1}{3}L$ $L = (3/2)\ \lambda_3$ $= (3/2)(0.340\text{ m})$ $L = 0.510$ m
Part b. Step 1. Draw an accurate diagram for the 3rd harmonic and determine the wavelength of the sound.	3rd harmonic $\vert\leftarrow$ L $\rightarrow\vert$ $\vert\leftarrow \frac{1}{4}\lambda_3 \rightarrow\vert$ $\frac{1}{4}\lambda_3 = \frac{1}{3}L$ $\lambda_3 = 3(\frac{1}{4})(0.510\text{ m})$ $\lambda_3 = 0.383$ m

Part b. Step 2. Determine the frequency of the sound waves.	$f = v/\lambda$ $= (340\ m/s)/(0.383\ m)$ $f = 888\ Hz$

TEXTBOOK QUESTION 7. How will the temperature in a room affect the pitch of organ pipes?

ANSWER: The frequency the first harmonic of an open pipe is given by $f_1 = v/(2L)$, while for a closed pipe $f_1 = v/(4L)$. Therefore, the frequency is directly proportional to the velocity of sound in air. The velocity of sound in air changes with temperature, i.e., $v = (331 + 0.60\ T)$ m/s where the temperature T is measured in degrees Celsius. As the temperature increases, the velocity increases and the pitch produced by the organ pipe increases.

Quality

The **quality** of a sound depends on the presence of overtones, their number, and relative amplitudes. A piano and a guitar may be playing the same note, with the same frequency and amplitude, yet the sounds are clearly distinguishable. The reason for this is that the relative amplitudes of the harmonics that are produced are different for different instruments and the note which is produced is the result of the superposition of the various harmonics. The result gives each instrument its characteristic quality and timbre.

Even two guitars will sound different because of certain characteristics in the construction of the instrument. These characteristics determine the relative amplitudes of the harmonics.

Music Versus Noise

Our minds interpret sounds that include frequencies that are simple multiples of one another as harmonious or pleasing to the ear. Noise is the result of many vibrations of different frequencies and amplitudes with no particular relationship to one another. Oftentimes, the distinction between music and noise is not sharp. The determination must be made by the individual. Thus what is music to the ears of the typical teenager might be considered noise by the typical parent.

Interference of Sound Waves

An **interference pattern** will be produced by two sources of sound waves separated by a certain distance (d) if the sounds produced are of the same frequency and amplitude. The interference pattern represented at the top of the next page is the result of sound produced by two sources, source 1 (s_1) and source 2 (s_2), both of which are in phase. The term "in phase" means that both sources produce compressions at the same moment of time and expansions at the same moment of time.

At locations where the path difference is a whole number multiple of the wavelength, the waves will arrive in phase and constructive interference occurs. Point A in the diagram is such a point. The distance from source 1 to point A is 1λ, while from source 2 the distance is 2λ. The path difference is therefore $2\lambda - 1\lambda = 1\lambda$.

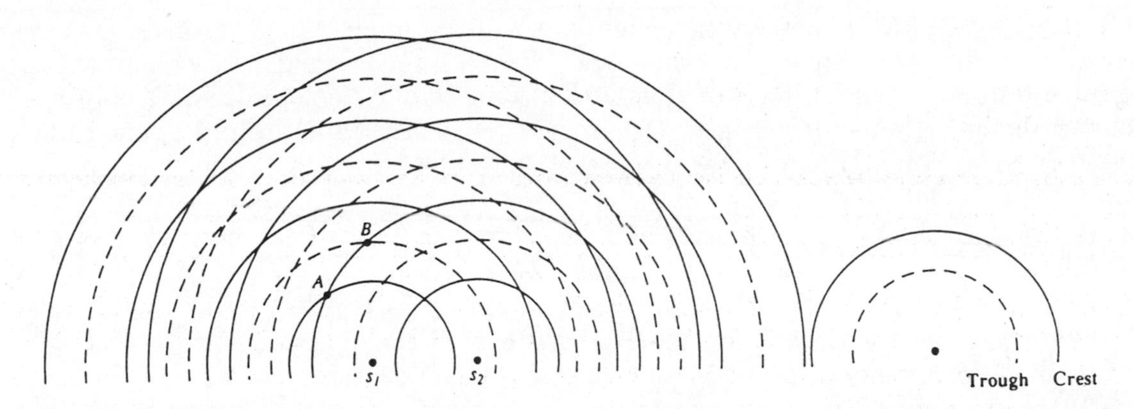

At locations where the path difference is odd multiples of ½λ, the waves arrive out of phase, i.e., a compression is superimposed on a expansion, and destructive interference occurs. Point B in the diagram is such a point. The distance from source 1 to point B is 3/2 λ and from source 2 is 2λ. The path difference is therefore $2\lambda - 3/2\,\lambda = ½\lambda$.

In the diagram shown below, the path difference is represented by the distance from source 1 to E. This distance is mλ for constructive interference and (m + ½)λ for destructive interference, where m = 0, 1, 2, 3, 4, etc.

The following formulas can be used to determine the positions of constructive and destructive interference if the angle θ is relatively small (less than 15°). For small angles, $\tan\theta \approx \sin\theta \approx \theta$, where θ is in radians.

For constructive interference: $\tan\theta = \Delta X/L$ but $\sin\theta = m\lambda/d$ and $\Delta X/L = m\lambda/d$

For destructive interference: $\tan\theta = \Delta X/L$ but $\sin\theta = (m + ½)\lambda/d$ and $\Delta X/L = (m + ½)\lambda/d$

ΔX is the perpendicular distance from the center line to the point in question. L is the length of the center line, mλ or (m + ½)λ is the path difference, and d is the distance between the sources of the sound waves.

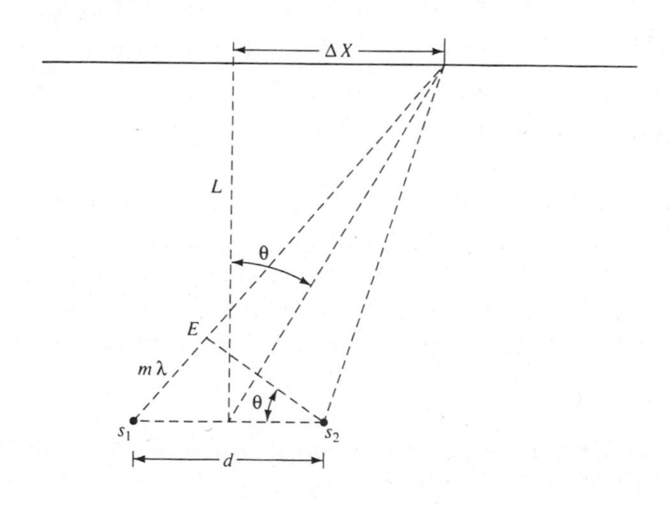

EXAMPLE PROBLEM 3. Two point sources of sound are in phase and separated by a distance (d) of 4.0 m. The frequency of the sound is 600 Hz. A listener stands at a point that is 10.0 m along a center line (L) that bisects the line which connects the two speakers. The listener then walks perpendicular to the center line. Determine the distance (ΔX) from the center line to the first two a) nodal points, b) antinodal points that are to the left of the center line. Assume the speed of sound is 340 m/s.

Part a. Step 1. Determine the wavelength of the sound waves.	Solution: (Section 12-7) $\lambda = v/f = (340\ m/s)/(600\ Hz)$ $\lambda = 0.567\ m$
Part a. Step 2. Draw an accurate diagram and list each of the quantities given.	$d = 4.0\ m$ $\quad$ $L = 10.0\ m$ $\lambda = 0.567\ m$ $\quad$ $\Delta X = ?$
Part a. Step 3. Apply the formula for destructive interference to determine the positions of the first two nodal points, m = 0 and m = 1.	first nodal point (m = 0); $\Delta X/L = (0 + ½)\lambda/d$ $\Delta X/(10.0\ m) = (½)(0.567\ m)/(4.0\ m)$ $\Delta X = 0.71\ m$ second nodal point (m = 1); $\Delta X/L = (1 + ½)\lambda/d$ $\Delta X/(10.0\ m) = (3/2)(0.567\ m)/(4.0\ m)$ $\Delta X = 2.13\ m$
Part b. Step 1. Apply the formula for constructive interference to determine the positions of the first two antinodal points to the left of the center line, m = 1 and m = 2.	The sources are in phase and therefore an antinodal point occurs along the center line (m = 0). The first two antinodal points to the the left of the center line occur at m = 1 and m = 2. $\Delta X/L = m\lambda/d$ where m = 1 for the 1st nodal point $\Delta X/(10.0\ m) = 1(0.567\ m)/(4.0\ m)$ $\Delta X = 1.42\ m$ $\Delta X/(10.0\ m) = 2(0.567\ m)/(4.0\ m)$ and m = 2 for 2nd nodal point: $\Delta X = 2.84\ m$

Beats

Two waves of equal amplitude but slightly different frequencies traveling in the same direction give rise to pulsations of maximum and minimum sound known as **beats**. The number of beats per second is equal to the difference between the frequencies of the component waves and is known as the beat frequency, $f_b = f_1 - f_2$. For example, if waves of 600 hertz and 610 hertz interfere to produce beats, the beat frequency would be 610 Hz - 600 Hz = 10 Hz, i.e., 10 beats per second would be heard.

EXAMPLE PROBLEM 4. A student strikes two tuning forks and hears 2 beats per second. He notes that 440 Hz is printed on one tuning fork. Determine the frequency of the other fork.

Part a. Step 1.

Determine the frequency of the second tuning fork.

Solution: (Section 12-7)

This problem has two possible answers. The difference between the two frequencies must be 2 Hz; however, we cannot be sure which tuning fork has the higher frequency.

Let f_1 = 440 Hz and f_2 = ?,

then either $f_1 - f_2 = 1$ Hz

440 Hz - f_2 = 2 Hz and f_2 = 438 Hz

or $f_2 - f_1 = 2$ Hz

f_2 - 440 Hz = 2 Hz and f_2 = 442 Hz

Therefore, the frequency of the second tuning fork may either be 438 Hz or 440 Hz.

Doppler Effect

When a source of sound waves and a listener approach one another, the pitch of the sound is increased as compared to the frequency heard if they remain at rest. If the source and the listener recede from one another, the frequency is decreased. This phenomena is known as the **Doppler effect**. The pitch heard by the listener is given by the following equations:

$f' = [1/(1 - v_s/v)]\, f$

source moving toward a stationary listener

$f' = [1/(1 + v_s/v)]\, f$

source moving away from a stationary listener

$f' = (1 + v_o/v)\, f$

listener moving toward a stationary source

$f' = (1 - v_o/v)\, f$

listener moving away from a stationary source

f' is the frequency of the sound heard by the listener (observer), f is the frequency of the sound emitted by the source, v is the speed of sound in air, v_s is the velocity of the source, and v_o is the velocity of the listener (observer).

Common examples of the Doppler effect include the change in pitch of the siren of a police car or ambulance as it passes at high speed, as well as the change in pitch of the whistle of a passing train.

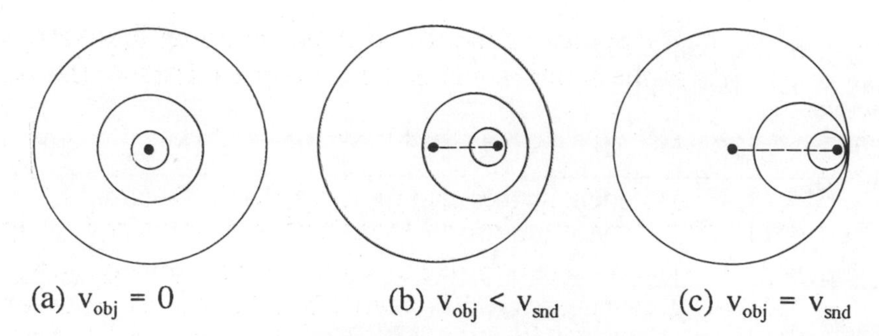

(a) $v_{obj} = 0$ (b) $v_{obj} < v_{snd}$ (c) $v_{obj} = v_{snd}$

Light waves also exhibit the Doppler effect. The spectra of stars that are receding from us is shifted toward the longer wavelengths of light. This is known as the **red shift**. Measurement of the red shift allows astronomers to calculate the speed at which stars are moving away. Since almost all stars and galaxies exhibit a red shift, it is believed that the universe is expanding.

TEXTBOOK QUESTION 17. Figure 12-32 shows the various positions of a child in motion on a swing. A monitor is blowing a whistle in front of the child on the ground. At which position will the child hear the highest frequency for the sound of the whistle? Explain your reasoning.

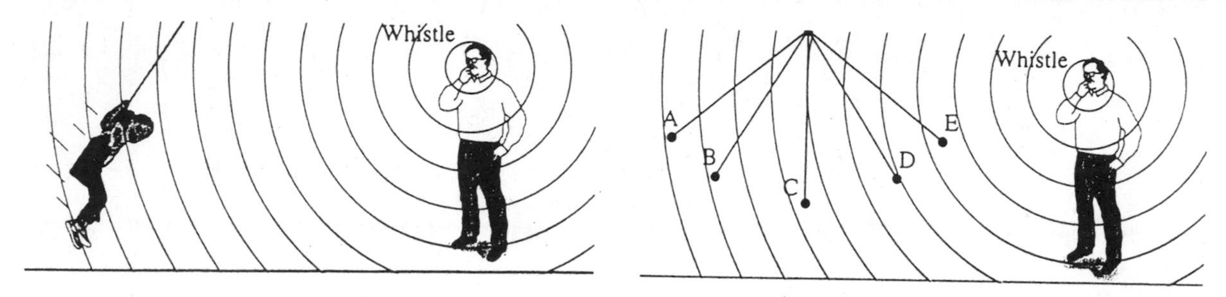

ANSWER: A child's speed on a swing is greatest at the bottom of the arc, i.e., at point C. The Doppler effect predicts that the frequency heard by the child will be greatest when the child's speed approaching the whistle is greatest. This occurs at point C as the child approaches the whistle. The lowest frequency heard by the child also occurs at point C as the child is traveling away from the whistle.

EXAMPLE PROBLEM 5. A stationary source emits sound of frequency 500 Hz on a day when the speed of sound is 340 m/s. A listener moves toward the source of sound at 40 m/s. Determine the a) frequency heard by the listener, b) wavelength of the sound between the source and the listener, and c) answer parts a and b if the listener was moving away from the source.

Part a. Step 1.	Solution: (Section 12-8)
Determine the frequency heard by the listener.	This problem involves the Doppler effect. Apply the formula for a moving listener approaching a stationary source. $f' = [1 + (v_o)/(v)]\ f$ $f' = [1 + (40\ m/s)/(340\ m/s)](500\ Hz) = 560\ Hz$

Part b. Step 1. Determine the wavelength of the sound between the source and the listener.	The source of the sound is not moving; therefore, the wavelength of the sound remains the same regardless of the listener's speed. $\lambda = v/f$ $\lambda = (340\ m/s)/(500\ Hz) = 0.68\ m$
Part c. Step 1. Solve for the listener moving away.	The wavelength of the sound will not change whether the listener is moving toward or away from the source; however, the frequency heard will change. Apply the formula for a listener moving away from the source. $f' = [1 - (v_o)/(v)]\ f = [1 - (40\ m/s)/(340\ m/s)]\ (500\ Hz)$ $f' = 441\ Hz$

Shock Waves and the Sonic Boom

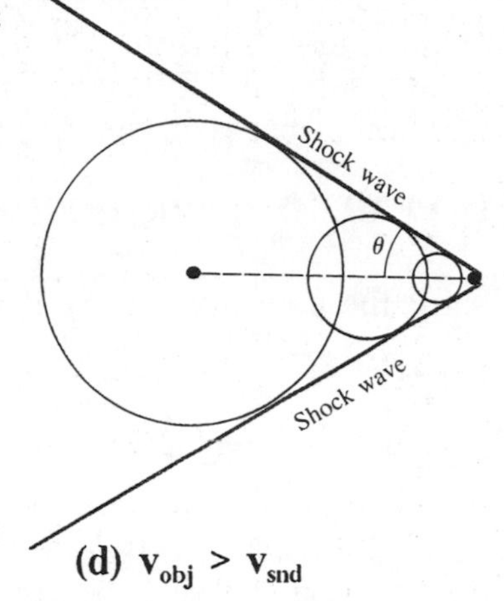

(d) $v_{obj} > v_{snd}$

When the speed of a source of sound exceeds the speed of sound, the sound waves in front of the source tend to overlap and constructively interfere. The superposition of the waves produce an extremely large amplitude wave called a **shock wave**.

The shock wave contains a great deal of energy. When the shock wave passes a listener, this energy is heard as a **sonic boom**. The sonic boom is heard only for a fraction of a second; however, it sounds as if an explosion has occurred and can cause damage.

PROBLEM SOLVING SKILLS

For problems involving the sound intensity:

1. Complete a data table noting the intensity level of the sound, the intensity of the threshold of hearing, frequency of the sound, the density of air, and the speed of sound.
2. Solve for the sound intensity in decibels and the amplitude of the sound wave.

For problems involving harmonics produced in pipes:

1. If necessary, solve for the speed of sound in the pipe.
2. Note whether the problem involves an open or closed pipe. Draw an accurate diagram locating the nodes and antinodes for the harmonic(s) requested.

3. Determine the wavelength of the waves producing the particular harmonic.
4. Solve for the frequency of the particular harmonic. If the frequency is given use the above steps to solve for the length of the pipe.

For problems involving an interference pattern produced by two sources of sound of the same frequency and amplitude which are separated by a distance d:

1. Draw an accurate diagram labeling d, ΔX, L, m, $m\lambda$, and λ.
2. Complete a data table listing the information given in the problem and the information requested in the solution.
3. Note whether the problem involves constructive interference or destructive interference.
4. Choose the appropriate formula(s) and solve the problem.

For problems involving the Doppler effect:

1. Complete a data table listing information both given and implied in the problem.
2. Note whether the source or the listener is moving; also note whether the source and listener are approaching each other or moving away from each other.
3. Select the appropriate formula and solve for frequency heard by the listener.
4. Solve for the wavelength of the sound between the source and the listener.

SOLUTIONS TO SELECTED TEXTBOOK PROBLEMS

TEXTBOOK PROBLEM 8. What is the intensity of sound at the pain level of 120 dB? Compare it to a whisper at 20 db.

Part a. Step 1. Apply the formula for intensity level in decibels and determine the intensity level for the 120 dB sound.	Solution: (Section 12-2) $\beta = 10 \log (I/I_o)$ where $I_o = 1.0 \times 10^{-12}$ W/m^2. 120 dB = $10 \log (I/1.0 \times 10^{-12}$ W/m$^2)$ $12 = \log (I/1.0 \times 10^{-12}$ W/m$^2)$ From algebra, if $A = \log B$, then $B = 10^A$; therefore, $I/1.0 \times 10^{-12}$ W/m$^2 = 10^{12}$ $I = (1.0 \times 10^{12})(1.0 \times 10^{-12}$ W/m$^2) = 1.0$ W/m^2
Part a. Step 2. Apply the same formula and determine the intensity level for the 20 dB sound.	$\beta = 10 \log (I/I_o)$ 20 dB = $10 \log (I/1.0 \times 10^{-12}$ W/m$^2)$ $2 = \log (I/1.0 \times 10^{-12}$ W/m$^2)$ $I/1.0 \times 10^{-12}$ W/m$^2 = 10^2$ $I = (1.0 \times 10^2)(1.0 \times 10^{-12}$ W/m$^2) = 1.0 \times 10^{-10}$ W/m^2

Part a. Step 3. Determine the ratio of the sound intensity levels.	$(1.0\ W/m^2)/(1.0 \times 10^{-10}\ W/m^2) = 1.0 \times 10^{10}$ The intensity level for the sound at the level of pain is 10 billion times greater than that of a whisper.

TEXTBOOK PROBLEM 20. What would be the sound level (in dB) of a sound wave in air that corresponds to a displacement amplitude of vibrating air molecules of 0.13 mm at 300 Hz?

Part a. Step 1. Use the equation from chapter 11 of the textbook to determine the intensity.	Solution: (Section 12-2) $I = 2\ \pi^2\ v\ \rho\ f^2\ x_o^2$ $I = 2\ \pi^2 (343\ m/s)(1.29\ kg/m^3)(300\ Hz)^2 (1.3 \times 10^{-4}\ m)^2$ $I = 13.28\ W/m^2$
Part a. Step 2. Solve for the intensity level in decibels.	$\beta = 10 \log (I/I_o)$ where $I_o = 1.0 \times 10^{-12}\ W/m^2$ $= 10 \log [(13.28\ W/m^2)/(1.0 \times 10^{-12}\ W/m^2)]$ $= 10 \log (1.328 \times 10^{13})$ $\beta = (10)(13.1) \approx 130$ dB

TEXTBOOK PROBLEM 25. An organ pipe is 112 cm long. What are the fundamental and first three audible overtones if the pipe is (a) closed at one end, and (b) open at both ends.

Part a. Step 1. Draw a diagram of the fundamental frequency of the pipe closed at one end.	Solution: (Section 12-5) First harmonic = fundamental $L = \frac{1}{4}\lambda_1$ $f_1 = v/(4L)$
Part a. Step 2. Determine the wavelength of the sound waves.	The distance between adjacent nodes and antinodes is $\frac{1}{4}\lambda$. Based on the diagram, the length of the pipe equals $\frac{1}{4}\lambda$. $1.12\ m = \frac{1}{4}\lambda$; therefore, $\lambda = 4.48$ m
Part a. Step 3. Determine the fundamental frequency of the waves. Assume that the speed of sound is 343 m/s.	$v = f\ \lambda$ $343\ m/s = f_1\ (4.48\ m)$ $f_1 = 76.6$ Hz

Part a. Step 4. Determine the wavelength of the first three audible overtones.	For a pipe closed at one end, the frequencies heard are consecutive odd multiples of the fundamental, i.e., 3, 5, 7, etc. times the fundamental frequency. $f_3 = 3(76.6 \text{ Hz}) = 230 \text{ Hz}$ $f_5 = 5(76.6 \text{ Hz}) = 383 \text{ Hz}$ $f_7 = 7(76.6 \text{ Hz}) = 536 \text{ Hz}$
Part b. Step 1. Draw a diagram of the fundamental frequency of the pipe open at both ends.	L First harmonic = fundamental $L = \frac{1}{2}\lambda_1$ $f_1 = v/(2L)$
Part b. Step 2. Determine the wavelength of the sound waves.	The distance between adjacent nodes is $\frac{1}{2}\lambda$. Based on the diagram, the length of the pipe equals $\frac{1}{2}\lambda$. $1.12 \text{ m} = \frac{1}{2}\lambda$; therefore, $\lambda = 2.24 \text{ m}$
Part a. Step 3. Determine the fundamental frequency of the sound.	$v = f\,\lambda$ $343 \text{ m/s} = f_1\,(2.24 \text{ m})$ $f_1 = 153 \text{ Hz}$
Part a. Step 4. Determine the wavelength of the first three audible overtones.	For a pipe open at both ends, the frequencies heard are consecutive whole number multiples of the fundamental, i.e., 2, 3, 4 etc. times the fundamental frequency. $f_2 = 2(153 \text{ Hz}) = 306 \text{ Hz}$ $f_3 = 3(153 \text{ Hz}) = 459 \text{ Hz}$ $f_4 = 4(153 \text{ Hz}) = 612 \text{ Hz}$

TEXTBOOK PROBLEM 46. Two loud speakers are 1.80 m apart. A person stands 3.00 m from one speaker and 3.50 m from the other speaker. (a) What is the lowest frequency at which destructive interference will occur at this point? (b) Calculate two other frequencies that also result in destructive interference at this point (give the next two highest). Let T = 20°C.

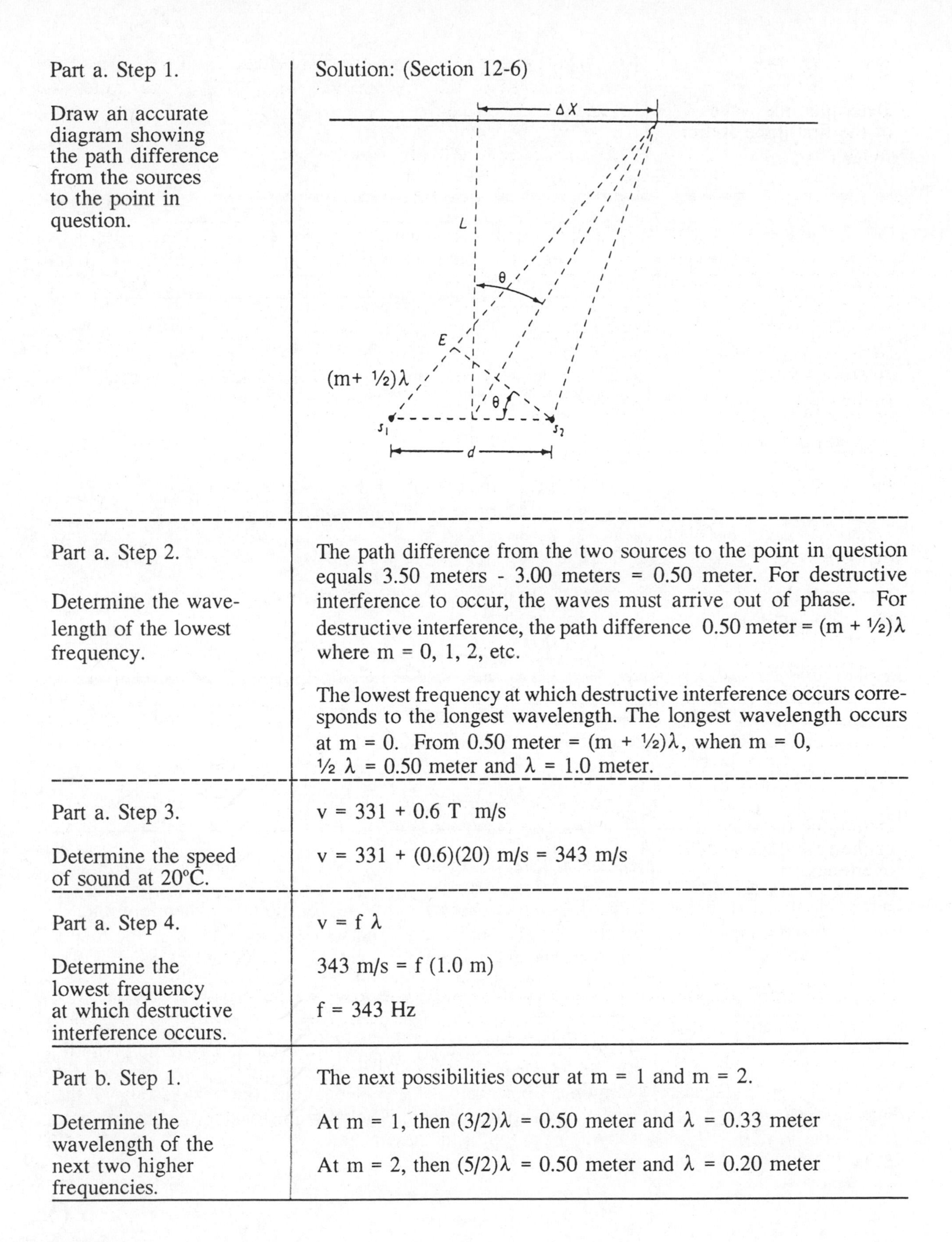

Part a. Step 1. Draw an accurate diagram showing the path difference from the sources to the point in question.	Solution: (Section 12-6) ΔX, L, θ, E, $(m + ½)\lambda$, θ, s_1, s_2, d
Part a. Step 2. Determine the wavelength of the lowest frequency.	The path difference from the two sources to the point in question equals 3.50 meters - 3.00 meters = 0.50 meter. For destructive interference to occur, the waves must arrive out of phase. For destructive interference, the path difference 0.50 meter = $(m + ½)\lambda$ where m = 0, 1, 2, etc. The lowest frequency at which destructive interference occurs corresponds to the longest wavelength. The longest wavelength occurs at m = 0. From 0.50 meter = $(m + ½)\lambda$, when m = 0, $½\ \lambda$ = 0.50 meter and λ = 1.0 meter.
Part a. Step 3. Determine the speed of sound at 20°C.	$v = 331 + 0.6\ T$ m/s $v = 331 + (0.6)(20)$ m/s = 343 m/s
Part a. Step 4. Determine the lowest frequency at which destructive interference occurs.	$v = f\ \lambda$ 343 m/s = f (1.0 m) f = 343 Hz
Part b. Step 1. Determine the wavelength of the next two higher frequencies.	The next possibilities occur at m = 1 and m = 2. At m = 1, then $(3/2)\lambda$ = 0.50 meter and λ = 0.33 meter At m = 2, then $(5/2)\lambda$ = 0.50 meter and λ = 0.20 meter

Part b. Step 2. Determine the next two higher frequencies.	$v = f \lambda$ 343 m/s = f (0.33 m), f = 1030 Hz 343 m/s = f (0.20 m), f = 1720 Hz

TEXTBOOK PROBLEM 49. The predominant frequency of a fire engine's siren is 1550 Hz when at rest. What frequency do you detect if you move with a speed of 30.0 m/s (a) toward the fire engine, and (b) away from the fore engine.

Part a. Step 1. Determine the frequency you hear as you approach the car.	Solution: (Section 12-7) This problem involves the Doppler effect. Apply the formula for a moving listener approaching a stationary source. $f' = [1 + (v_o)/(v)]f$ $f' = [1 + (30\ m/s)/(343\ m/s)](1550\ Hz) = 1690\ Hz$
Part b. Step 1. Determine the frequency heard as you recede.	$f' = [1 - (v_o)/(v)]\ f$ $f' = [1 - (30\ m/s)/(343\ m/s)](1550\ Hz)$ $f' = 1410\ Hz$

TEXTBOOK PROBLEM 76. A tuning fork is set into vibration above a vertical open tube filled with water (Fig. 12-35). The water level is allowed to drop slowly . As it does so, the air in the tube above the water level is heard to resonate with the tuning fork when the distance from the tube opening to the water level is 0.125 m and again at 0.395 m. What is the frequency of the tuning fork?

Part a. Step 1. Determine whether the tube acts like a pipe closed at one end or open at both ends.	Solution: (Section 12-5) Sound is free to pass out of the top of the tube. Therefore, the top of the tube is an open end. The sound reflects back up the tube at the water level. Therefore, the water level acts as a closed end. The tube is a pipe closed at one end.
Part a. Step 2 Draw an accurate diagram for the water level at the 1st harmonic and the 3rd harmonic.	The water level is slowly lowered from the top of the pipe until the first resonance occurs. This resonance represents the 1st harmonic. For a pipe closed at one end, the next resonance that occurs is the 3rd harmonic. The diagram for the third harmonic is shown at the top of the next page.

	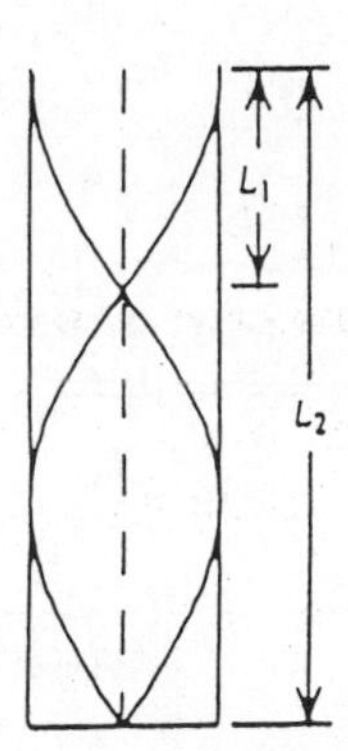
Part a. Step 3. Use the diagrams in step 3 to determine the wavelength of the sound waves.	At the 1st harmonic, $L_1 = \frac{1}{4}\lambda$. At the 3rd harmonic $L_3 = \frac{3}{4}\lambda$. The distance between the node of the 1st harmonic and the node of the 3rd harmonic $\Delta L = L_3 - L_1 = 0.395\ m - 0.125\ m = 0.270\ m$. However, as shown in the diagram, $\Delta L = \frac{3}{4}\lambda - \frac{1}{4}\lambda = \frac{1}{2}\lambda$. $\frac{1}{2}\lambda = 0.270\ m$ $\lambda = 0.540\ m$
Part a. Step 4. Determine the frequency of the tuning fork.	$f = v/\lambda$ $f = (343\ m/s)/(0.540\ m) = 635\ hz$

CHAPTER 13

TEMPERATURE AND KINETIC THEORY

OBJECTIVES

After studying the material of this chapter, the student should be able to:

- convert a temperature given in degrees Fahrenheit to degrees Celsius and/or degrees Kelvin, and vice versa.
- state the factors that cause the volume of a solid or liquid to change or the length of a solid to change. Also, solve word problems and determine the final length or volume.
- write the mathematical relationships that summarize Boyle's law, Charles's law, Gay-Lussac's law, and the ideal gas equation. Use these equations to solve word problems.
- state in your own words Avogadro's hypothesis. State from memory the modern value of Avogadro's number.
- state the postulates of the kinetic theory of gases.
- rewrite the ideal gas equation in terms of motion of the molecules of an ideal gas.
- explain what is meant by the term rms velocity.
- given a phase diagram for water, determine the range of temperature and pressure at which water is a solid, liquid, or gas. Describe what is meant by the triple point of water and point out the triple point on a phase diagram.
- explain what is meant by sublimation and use a phase diagram to determine the range of temperatures and pressures for which the sublimation of water could occur.
- explain why evaporation from a liquid is related to the temperature of the liquid and the average kinetic energy of the molecules of the liquid.
- explain what is meant by vapor pressure and explain why vapor pressure is related to the temperature of the liquid and the boiling point of the liquid.
- distinguish between relative humidity and absolute humidity and solve word problems related to relative humidity.
- explain what is meant by diffusion and why diffusion is slower through a liquid than through a gas.
- use Fick's law to solve word problems related to gaseous diffusion.
- state Graham's law of diffusion and use this law to determine the mass of a molecule of an unknown gas.

KEY TERMS AND PHRASES

temperature is a measure of relative hotness or coldness. In kinetic theory, temperature is a measure of the average kinetic energy of the particles of a substance.

thermometer is a device used to measure the temperature of a substance. The thermometer is usually calibrated using either the Fahrenheit temperature scale (°F) or the Celsius (°C). Water is the reference used by both temperature scales. At 1 atmosphere of pressure, the freezing point of water is 32°F or 0°C, while the boiling point is 212°F or 100°C.

coefficient of linear expansion is the ratio between the change in length of a solid rod to its original length per one degree change in temperature.

coefficient of volume expansion is the ratio between the change in volume of a solid or a liquid to its original volume per one degree change in temperature.

Boyle's law states that, if the temperature of a gas is held constant, the volume occupied by an enclosed gas varies inversely with the pressure exerted on it.

Charles's law states that the volume of an enclosed gas held at constant pressure is directly proportional to the absolute temperature.

Gay-Lussac's law states that the pressure exerted by a gas held at constant volume is directly proportional to the absolute temperature.

ideal gas refers to a monatomic, low density and low pressure gas. The idea of an 'ideal' or perfect gas is used to create a model which can then be used to study real gases.

ideal gas law states that the three gas laws, Boyle's, Charles's, and Gay-Lussac's, can be combined into a single more general law that relates the pressure (P), volume (V), and temperature (T) of a fixed quantity of gas (n).

Avogadro's hypothesis states that equal volumes of gas at the same pressure and temperature contain equal numbers of molecules.

Avogadro's number is the number of atoms (or molecules) per mole. The modern value of Avogadro's number is 6.02×10^{23} molecules.

postulates of the kinetic theory of gases are 1) a gas consists of a large number of molecules moving in random directions with a variety of speeds, 2) the average distance between any two molecules in a gas is large compared to the size of an individual molecule, 3) the molecules obey the laws of classical mechanics and are presumed to interact with one another only when they collide, 4) collisions between molecules or between a molecule and the side walls of the container are perfectly elastic.

root-mean-square velocity (v_{rms}) refers to the square root of the average of the squares of the magnitudes of the velocities of the molecules in a gas.

phase diagram is a pressure-temperature diagram that allows the reader to determine the temperatures and pressures at which the various phases (solid, liquid, gas) for a particular substance exist. The diagram also indicates those conditions under which equilibrium occurs.

triple point is the point at which all three states of matter (solid, liquid, gas) coexist in equilibrium. For water, the triple point is at 0.01°C and 0.006 atm.

sublimation is the direct conversion of a solid to a vapor without passing through the liquid state.

evaporation is the process by which molecules escape from a liquid and enter the gas or vapor phase.

vapor pressure refers to the partial pressure contributed by the molecules that have evaporated from a liquid.

relative humidity is a measure of the water vapor content of the air compared to the maximum amount the air can hold at a particular temperature.

dew point is reached when the relative humidity reaches the saturation point and the air can hold no more water. If the temperature drops below this point, condensation or dew occurs.

boiling occurs when the temperature of the liquid rises to the point where the vapor pressure inside the bubbles that form in the liquid equals or exceeds the external air pressure.

diffusion is the spontaneous movement of molecules from a region of high concentration to a region of low concentration due to random molecular movement. The rate of diffusion of one substance into another is described by the **Fick's law.**

Graham's law describes the rate of diffusion of gas molecules. The law states that the rate of diffusion of molecules of a gas is inversely proportional to the square root of the molecular mass.

SUMMARY OF MATHEMATICAL FORMULAS

Celsius-Fahrenheit temperature conversion	$°F = 9/5\ °C + 32°$ or $°C = 5/9\ (°F - 32°)$	These formulas allow the conversion of temperatures from Fahrenheit to Celsius and vice versa.
linear expansion	$\Delta L = L_o\ \alpha\ \Delta T$ and $L = L_o\ (1 + \alpha\ \Delta T)$	The linear expansion (ΔL) of a solid rod is related to the rod's original length (L_o), coefficient of linear expansion (α) and the change in temperature (ΔT).
volume expansion	$\Delta V = V_o\ \beta\ \Delta T$ and $V = V_o\ (1 + \beta\ \Delta T)$	The volume expansion (ΔV) of a substance is related to the original volume (V_o), coefficient of volume expansion (β), and the change in temperature (ΔT).
Boyle's law	$P\ V = \text{constant}$ $(P_1)(V_1) = (P_2)(V_2)$	If the temperature of a gas is held constant, the volume occupied by an enclosed gas varies inversely with the pressure exerted on it.
Charles's law	$V/T = \text{constant}$ $(V_1)/(T_1) = (V_2)/(T_2)$	The volume of an enclosed gas held at constant pressure is directly proportional to the absolute temperature.

Gay-Lussac's law	P/T = constant $(P_1)/(T_1) = (P_2)/(T_2)$	The pressure exerted by a gas held at constant volume is directly proportional to the absolute temperature.
ideal gas law	$P V = n R T$	The ideal gas law relates the pressure (P), volume (V), temperature (T), and the number of moles (n) in the gas. R is a constant of proportionality called the universal gas constant.
equations from kinetic theory of gases	$P V = \frac{1}{3} N m v^2$ $v_{rms} = (\overline{v^2})^{1/2}$ $\overline{KE} = 3/2\ k T$	N is the total number of molecules in the gas and m is the mass of an individual molecule. v_{rms} is the square root of average value of the squares of the velocities of the molecules. T is the temperature in degrees Kelvin. k is Boltzmann's constant.
relative humidity	$R.H. = (\rho_{wv})/(\rho_{swv}) \times 100\%$ $R.H. = (P_{vp})/(P_{svp}) \times 100\%$	The relative humidity (R.H.) is the ratio of the density of water vapor in the air (ρ_{wv}) as compared to the density of water vapor in the air when the air is saturated(ρ_{swv}). or The relative humidity is the ratio of the partial pressure of the water vapor in the air (P_{vp}) to the vapor pressure of water when the air is saturated (P_{svp}).
Diffusion Fick's law	$J = D A (C_1 - C_2)/d$	The rate of diffusion (J) of one substance into another is related to the diffusion constant (D), the cross-sectional area (A) of the region through which the diffusion occurs, and the concentration of the diffusing substance (C). C_1 is the region of higher concentration and C_2 is the region of lower concentration. Δx is the unit of distance .
gaseous diffusion Graham's law	$J \propto 1/(m)^{1/2}$	The rate of diffusion (J) of gas molecules is inversely proportional to the square root of the molecular mass.

CONCEPTS AND EQUATIONS

Temperature

Temperature is a measure of relative hotness or coldness. The temperature of a substance is measured with a device called a **thermometer**. The thermometer is usually calibrated in terms of degrees **Fahrenheit** (°F) or degrees **Celsius** (°C). Water is the reference used by both temperature scales. At 1 atmosphere of pressure, the freezing point of water is 32°F or 0°C, while the boiling point is 212°F or 100°C. Although the Fahrenheit temperature scale shall not be used in this study guide, it is useful to memorize the conversion formulas for Fahrenheit to centigrade or vice versa:

$°F = 9/5\ °C + 32°$ or $°C = 5/9\ (°F - 32°)$

In many instances it will be necessary to use the Kelvin temperature scale, where $K = °C + 273°$. The Kelvin temperature is also known as the absolute temperature scale.

Linear Expansion

The change in length (ΔL) of a solid when it is heated or cooled depends on the original length (L_o), coefficient of linear expansion (α), and the change in temperature (ΔT). The unit used for the coefficient of linear expansion is (/C°) or $(C°)^{-1}$. The change in length is given by $\Delta L = L_o\ \alpha\ \Delta T$. The final length (L) of the solid after the temperature change is given by

$$L = L_o + \Delta L = L_o (1 + \alpha\ \Delta T)$$

TEXTBOOK QUESTION 5. A flat bimetallic strip consists of aluminum riveted to a strip of iron. When heated, which metal will be on the outside of the curve? *Hint*: See Table 13-1. Why?

ANSWER: As shown in Table 13-1 in the textbook, the coefficient of linear expansion of aluminum is $25 \times 10^{-6}/C°$ while for iron the value is $12 \times 10^{-6}/C°$. When heated the aluminum will expand more than the iron. As a result, the aluminum will be on the outside of the curve as the strip bends.

TEXTBOOK QUESTION 14. Will a grandfather's clock, accurate at 20°C, run fast or slow on a hot day (30°C)? Explain. The clock uses a pendulum supported on a long, thin brass rod.

ANSWER: As the temperature increases the length of the pendulum increases ($\Delta L = L_o\ \alpha\ \Delta T$). As the length increases the period (T) of the pendulum increases, i.e., $T = 2\pi(L/g)^{1/2}$. As a result, it will take longer for the pendulum's motion to repeat and the clock will run slow.

Volume Expansion

The change in volume (ΔV) of a solid or liquid when it is heated or cooled depends on the original volume (V_o), coefficient of volume expansion (β), and the change in temperature (ΔT). The volume and/or change of volume of a substance is given in cm^3, m^3, or liters. The unit used for the coefficient of volume expansion is (/C°) or $(C°)^{-1}$. The change in volume (ΔV) of a solid or liquid that undergoes a temperature change is given by $\Delta V = V_o \beta \Delta T$, while the volume (V) at the final temperature is given by

$$V = V_o + \Delta V = V_o (1 + \beta \Delta T)$$

The Ideal Gas

The changes in volume of a gas cannot be described by the equation given for a solid or liquid. However, within a certain range of temperatures and pressures many gases have been found to follow three simple laws: Boyle's law, Charles's law and Gay-Lussac's law. A gas that follows these laws completely is an idealization called an **ideal gas**.

The variables involved in these laws are the pressure (P), the volume (V), the temperature (T), and the number of moles of gas (n). The Kelvin temperature scale is always used. One mole contains 6.02×10^{23} molecules and equals the molecular weight of the substance expressed in grams.

Boyle's Law

If the temperature of a gas is held constant, the volume occupied by an enclosed gas varies inversely with the pressure exerted on it. **Boyle's** law can be written as follows:

$V \propto 1/P$ and $P V = \text{constant}$ (at constant temperature)

Charles's Law

The volume of an enclosed gas held at constant pressure is directly proportional to the absolute temperature:

$V \propto T$ and $V/T = \text{constant}$ (at constant pressure)

Gay-Lussac's Law

The pressure exerted by a gas held at constant volume is directly proportional to the absolute temperature:

$P \propto T$ and $P/T = \text{constant}$ (at constant volume)

Ideal Gas Law

The three gas laws can be combined into a single more general law that relates the pressure (P), volume (V), and temperature (T) of a fixed quantity of gas (n):

$$P V = n R T$$

R is a constant of proportionality called the universal gas constant. Depending on the situation, R may be expressed in different sets of units. Typical values of R are

$$R = 8.31 \text{ J/(mol·K)} = 0.0821 \text{ (L atm)/(mol·K)} = 1.99 \text{ calories/(mol·K)} = 62.4 \text{ (liter torr)/(mol·K)}$$

EXAMPLE PROBLEM 1. An ideal gas occupies a volume of 1.50 liters at 303 K and gauge pressure of 1.00 atm. The gas is heated until its volume is 2.50 liters and gauge pressure is 1.20 atm. Determine a) the number of moles contained in the gas and b) the final temperature of the gas.

Part a. Step 1. Convert from gauge pressure to absolute pressure.	Solution: (Sections 13-7 and 13-8) absolute pressure = gauge pressure + 1.00 atm = 1.00 atm + 1.00 atm absolute pressure = 2.00 atm
Part a. Step 2. Determine the number of moles in the gas.	Apply the ideal gas law using the initial conditions to determine the number of moles in the gas. $P_i V_i = n R T_i$ (2.00 atm)(1.50 liter) = n [0.0821 (liter atm)/(mol·K)](303 K) n = 0.121 moles
Part b. Step 1. Determine the absolute pressure after the change.	The absolute pressure after the change has occurred is 1.20 atm + 1.00 atm = 2.20 atm.
Part b. Step 2. Use the ideal gas equation and solve for the final temperature.	Again apply the ideal gas formula and solve for the final temperature. $P_f V_f = n R T_f$ (2.20 atm)(2.50 liters) = (0.121 mole)[0.0821 (liter atm)/(mole K)]T_f T_f = 554 K

Avogadro's Number and the Ideal Gas Law

Avogadro's hypothesis formulated in 1811 states that "equal volumes of gas at the same pressure and temperature contain equal numbers of molecules." The number of molecules per mole is known as **Avogadro's number** (N_A). The modern value of N_A is 6.02×10^{23} molecules.

The total number of molecules in a gas (N) equals the product of the number of moles of gas (n) and the number of molecules per mole (N_A); thus $N = n N_A$ and the ideal gas equation can be written as follows:

$P V = n R T = (N/N_A) R T$ but $(N/N_A)R = k$

where $k = 1.38 \times 10^{-23}$ J/K and k is known as **Boltzmann's constant.**

Postulates of the Kinetic Theory of Gases

1) A gas consists of a large number of molecules moving in random directions with a variety of speeds.
2) The average distance between any two molecules in a gas is large compared to the size of an individual molecule.
3) The molecules obey the laws of classical mechanics and are presumed to interact with one another only when they collide.
4) Collisions between molecules or between a molecule and the side walls of the container are perfectly elastic.

The pressure exerted by an enclosed gas may be rewritten in light of these postulates and is given by the equation $P V = \frac{1}{3} N m v^2$ where N is the total number of molecules in the gas, m is the mass of an individual molecule, and v^2 is the average value of the squares of the velocities of the molecules of the gas.

Root-Mean-Square Velocity **(v_{rms})**

The root-mean-square velocity is often confused with average velocity. The root-mean-square velocity refers to the square root of the average of the squares of the magnitudes of the velocities of the molecules in a gas.

$$v_{rms} = (\overline{v^2})^{1/2} = [(v_1^2 + v_2^2 + ... + v_N^2)/N]^{1/2}$$

The average or the mean speed ($\bar{v}$) is equal to the sum of the speeds of the molecules divided by the number of molecules.

$$\bar{v} = (v_1 + v_2 + ... + v_N)/N$$

EXAMPLE PROBLEM 2. Five molecules have the following speeds: 3.1 m/s, 4.3 m/s, 5.6 m/s, 6.2 m/s, 8.5 m/s. Determine the a) mean speed and b) rms speed.

Part a. Step 1. Determine the mean speed.	Solution: (Section 13-10) The mean speed ($\bar{v}$) equals the sum of the individual speeds divided by the total number of molecules. $\bar{v} = [3.1\ m/s + 4.3\ m/s + 5.6\ m/s + 6.2\ m/s + 8.5\ m/s]/5$ $\bar{v} = 5.5\ m/s$
Part b. Step 1. Determine the rms speed.	The rms speed is equal to the square root of the mean of the square of the speeds of the molecules of the gas. $v_{rms} = [(3.1)^2 + (4.3)^2 + (5.6)^2 + (6.2)^2 + (8.5)]/5]^{1/2}$ $v_{rms} = 5.8\ m/s$

EXAMPLE PROBLEM 3. A 2.00 liter vessel contains 0.0300 kg of an ideal gas at 300°C and a pressure of 5.00 atm. a) How many moles of gas are in the vessel? b) Determine the root-mean-square speed of the molecules in the gas.

Part a. Step 1.	Solution: (Sections 13-7, 13-8 and 13-10)
Apply the ideal gas equation and solve for the number of moles of gas present.	Express the temperature in degrees Kelvin: $300°C + 273°C = 573$ K $P\ V = n\ R\ T$ $(5.00\ \text{atm})(2.00\ \text{liter}) = n\ [0.0821\ (\text{liter atm})/(\text{mol}\cdot\text{K})](573\ \text{K})$ $n = 0.213$ moles
Part b. Step 1.	$n = (0.213\ \text{mole})(6.02 \times 10^{23}\ \text{molecules/mole})$
Determine the number of molecules present.	$n = 1.28 \times 10^{23}$ molecules
Part b. Step 2.	$m = (0.0300\ \text{kg})/(1.28 \times 10^{23}\ \text{molecules})$
Determine the mass of a single molecule.	$m = 2.34 \times 10^{-25}$ kg/molecule
Part b. Step 3.	$\overline{KE} = (3/2)\ k\ T$ and $\frac{1}{2}\ m\ \overline{v^2} = 3/2\ k\ T$
Determine the rms speed.	$v_{rms} = (3\ k\ T/m)^{1/2}$ $= [3\ (1.38 \times 10^{-23}\ \text{J/molecule K})(573\ \text{K})/(2.34 \times 10^{-25}\ \text{kg})]^{1/2}$ $v_{rms} = 318$ m/s

Relationship between Average Kinetic Energy and the Absolute Temperature

From the kinetic theory, $P\ V = \frac{1}{3}\ N\ m\ \overline{v^2}$ and from the ideal gas law, $P\ V = n\ R\ T$. If we express both equations in the same units, then

$\frac{1}{3}\ N\ m\ \overline{v^2} = n\ R\ T$ and rearranging we find

$\frac{1}{2}\ m\ \overline{v^2} = \frac{1}{2}\ (3\ n\ R\ T/N) = (3/2)(nR/N)\ T$ but $nR/N = k =$ constant, therefore

$\overline{KE} = (3/2)\ k\ T$

From this equation we can see that the average kinetic energy of a molecule in a gas is directly proportional to the absolute temperature of the gas.

EXAMPLE PROBLEM 4. Calculate the a) average translational kinetic energy and b) rms speed of a neon atom when the temperature is 20.0°C. Note: neon is a monatomic gas and one mole of neon has a mass of 20.2 grams.

Part a. Step 1.	Solution: (Sections 13-10, 13-11)
Convert 20.0°C to degrees kelvin.	$20.0°C + 273°C = 293\ K$
Part a. Step 2. Determine the translational KE of a neon atom.	$\overline{KE} = 3/2\ k\ T = (3/2)(1.38 \times 10^{-23}\ J/K)(293\ K)$ $\overline{KE} = 6.07 \times 10^{-21}\ J$
Part b. Step 1. Determine the mass m in kg of a neon atom.	$m = (20.2\ g/mole)(1.0\ kg/1000\ g)(1\ mole/6.02 \times 10^{23}\ atoms)$ $m = 3.36 \times 10^{-26}\ kg$
Part b. Step 2. Calculate the rms speed of a neon atom.	$\overline{KE} = ½\ m\ v^2$ but $\overline{KE} = 3/2\ k\ T$ $½\ (3.36 \times 10^{-26}\ kg)\ \overline{v^2} = 3/2\ (1.38 \times 10^{-23}\ J/K)(293\ K)$ $\overline{v^2} = 3.61 \times 10^5\ m^2/s^2$ $v_{rms} = 601\ m/s$

Real Gases and Changes of Phase

In Section 13-11 of the textbook, the second postulate of the kinetic theory assumes that the volume occupied by the gas molecules is negligible compared to the volume of the container. However, when dealing with a real gas a correction factor must be included. Especially at high pressures, when the volume occupied by the gas is small, the behavior of a real gas deviates significantly from the predictions of the ideal gas law.

The third postulate states that "molecules of a gas are presumed to interact with one another only when they collide." However, contrary to this postulate, all molecules, even supposedly inert molecules such as helium, exhibit weak, short-range attractive forces for one another. These forces, called Van der Waal forces, are ignored when discussing the ideal gas law. If sufficient pressure is applied, it is found that as a gas cools the molecules begin to cling together and the gas will become a liquid below a certain **critical temperature**. For helium, this temperature is about -267°C (6 K). Above the critical temperature, the gas cannot be liquified, no matter what the pressure. Diagram A (at the top of the next page) shows a possible variation of volume with temperature for a real gas as compared to the predictions from the ideal gas law.

A pressure-temperature diagram (**phase diagram**) allows us to determine the temperatures and pressures at which the various phases (solid, liquid, gas) exist, as well as those conditions under which equilibrium occurs.

Diagram B, shown below, is the phase diagram for water. The point where the three lines intersect is called the **triple point**. At this point, all three states of matter (solid, liquid, and gas) coexist in equilibrium. For water, the triple point is at 0.01°C and 0.006 atm.

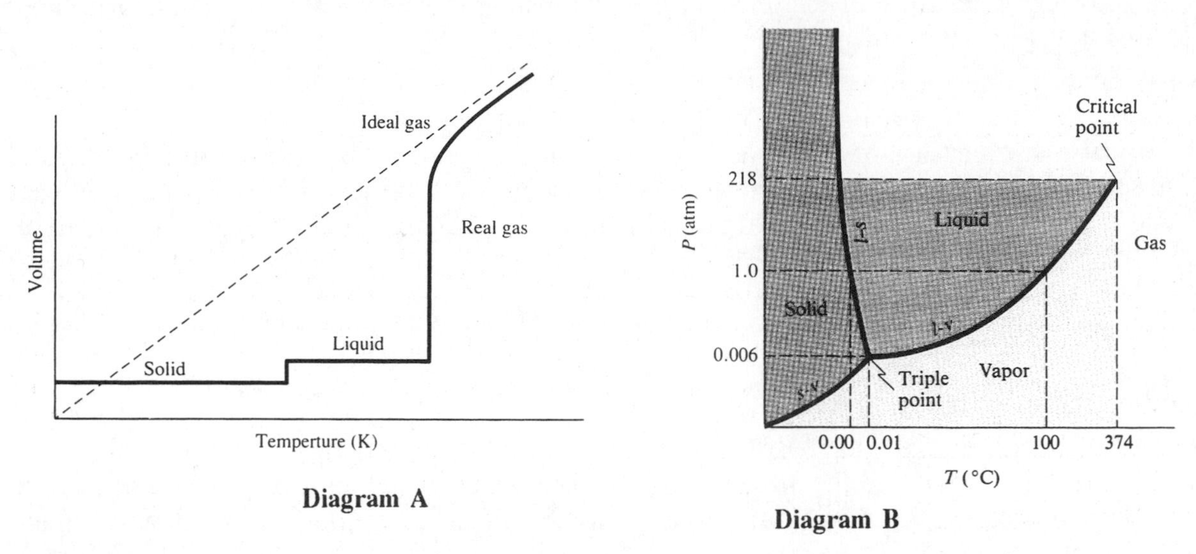

Diagram A

Diagram B

Sublimation

Sublimation is the direct conversion of a solid to a vapor without passing through the liquid state. As shown in the Diagram B, the sublimation of ice to the vapor phase does not occur at ordinary temperatures and pressures. For water, the temperature must be below 0.01°C and the pressure must be below 0.006 atm. However, common substances that exhibit sublimation at room temperature and pressure are moth balls (naphthalene) and dry ice (CO_2). They go directly from the solid phase to the vapor phase.

Evaporation

Evaporation is the process by which molecules escape from a liquid and enter the gas or vapor phase. The molecules that escape overcome forces which hold them in the liquid phase. It is the molecules of high kinetic energy that leave the liquid, with the result that the average kinetic energy of the remaining molecules decreases. Since temperature is proportional to the average kinetic energy of the molecules, the temperature of the liquid decreases. Thus evaporation is a cooling process.

If the temperature of the liquid increases, it is observed that the evaporation rate increases. The kinetic theory predicts that at the higher temperature, more molecules will have kinetic energy in excess of that required to escape from the liquid and is therefore in agreement with observation.

Vapor Pressure

The molecules that have evaporated from the liquid travel in random directions. Some of these molecules return to the liquid but evaporation will continue until the number of molecules leaving the liquid per second equals the number that are reentering the liquid per second. This point is the equilibrium point and the air is then said to be saturated.

The evaporated molecules contribute to the total pressure on the liquid from which they escaped. The **partial pressure** due to these molecules after (saturation) equilibrium has been reached is called the **vapor pressure**. At higher temperatures the vapor pressure is higher due to the greater number of molecules that have escaped from the liquid. At the boiling point of the liquid, the vapor pressure equals the atmospheric pressure. Table 13-3 in the textbook lists the saturated vapor pressure of ice and water at various temperatures. Ice is included because solids also have a very small, but measurable, vapor pressure.

Humidity

Humidity is a measure of the water vapor content of the air. **Absolute humidity** refers to the mass of water vapor present per unit volume of gas. When discussing atmospheric conditions, the term **relative humidity** (R.H.) is used. Relative humidity is determined as follows:

R.H. = [density of water vapor in the air]/[density of saturated air at the same temperature] x 100%

or R.H. = [partial pressure of water vapor]/[saturated vapor pressure of water] x 100%

If the temperature of the air begins to drop, e.g., after sundown, the relative humidity may reach the saturation point and the air can hold no more water. If the temperature begins to drop below this point, called the **dew point, condensation** will occur in the form of dew on grass, cars, etc. or possibly as fog.

TEXTBOOK QUESTION 23. Explain why a hot, humid day is far more uncomfortable than a hot, dry day at the same temperature.

ANSWER: On a hot day the human body perspires and uses evaporative cooling to keep cool. However, the partial pressure of water in the air on a hot, humid day is much higher than on a hot, dry day. As a result, the rate of evaporative cooling of perspiration on a humid day is lower and droplets of perspiration tend to form and the person feels warm, "sticky," and uncomfortable. On a hot, dry day perspiration evaporates faster and the person feels cooler.

EXAMPLE PROBLEM 5. A cool mist vaporizer is used in the bedroom of a child who has a common cold. The relative humidity of the room is 20.0% at 20.0°C. The room is 3.00 m x 3.00 m x 2.70 m. Determine the amount of water vapor in grams and liters that must be added to the air of the room in order to raise the relative humidity to 60.0%.

Part a. Step 1.	Solution: (Section 13-13)
Determine the initial partial pressure of the water vapor in torr.	According to Table 13-3 the saturated vapor pressure of water at 20.0°C is 17.5 torr. R.H. = $(P_p)/(P_v) \times 100\%$ $20.0\% = (P_p)/17.5 \text{ torr} \times 100\%$

	$P_p = (20.0\%)/(100\%) \times 17.5$ torr $= 3.50$ torr
Part a. Step 2. Determine the number of moles of water vapor present at 20.0°C. Note: 20.0°C = 293 K	$V = 3.00 \text{ m} \times 3.00 \text{ m} \times 2.70 \text{ m} = 24.3 \text{ m}^3$ $P V = n R T$ $(3.50 \text{ torr})(24.3 \text{ m}^3)(10^3 \text{ liter}/1 \text{ m}^3) = n\ (62.4 \text{ liter torr/mol}\cdot\text{K})(293 \text{ K})$ $n = 4.65$ moles
Part a. Step 3. Determine the number of grams of water in 4.65 moles.	A water molecule contains 2 atoms of hydrogen and 1 atom of oxygen. The molecular weight of water is $(2 \times 1) + (1 \times 16) = 18.0$ g/mole. The number of grams of water added is 4.65 moles x 18.0 grams/mole = 83.7 grams
Part a. Step 4. Determine the volume of water initially present in the vapor phase.	The density of water is 1.00 g/ml and there are 1000 ml in 1.00 liter. $V = (83.7 \text{ g})(1 \text{ ml/g})(1 \text{ liter}/1000 \text{ ml}) = 0.0837$ liters
Part a. Step 5. Determine the partial pressure of the water vapor in torr when the relative humidity is 60.0%.	$R.H. = (P_p)/(P_v) \times 100\%$ $60.0\% = (P_p)/17.5 \text{ torr} \times 100\%$ $P_p = (60.0\%)/(100\%) \times 17.5$ torr $P_p = 10.5$ torr
Part a. Step 6. Determine the number of moles of water vapor present at 20.0°C when the R.H. = 60.0%.	$P V = n R T$ $(10.5 \text{ torr})(24.3 \text{ m}^3)(10^3 \text{ liter}/1 \text{ m}^3) = n\ (62.4 \text{ liter torr/mol}\cdot\text{K})(293 \text{ K})$ $n = 14.0$ moles
Part a. Step 7. Determine the number of grams of water in 14.0 moles.	14.0 moles x 18.0 grams/mole = 252 grams

Part a. Step 8. Calculate the number of grams of water needed to raise the R.H. from 20.0% to 60.0%.	252 g required - 83.7 g initially present = 168 g to be added
Part a. Step 9. Determine the volume of water added to the vapor phase.	The density of water is 1.00 g/ml and there are 1000 ml in 1.00 liter. V = (168 g)(1 ml/gram)(1.00 liter/1000 ml) = 0.168 liters

Boiling

Boiling occurs when the temperature of the liquid rises to the point where the vapor pressure inside the bubbles that form in the liquid equals or exceeds the external air pressure. Archimedes principle predicts that when the buoyant force exerted on the bubble by the liquid exceeds the weight of the bubble, the bubble will rise to the surface.

The boiling point of a liquid depends on the atmospheric (barometric) pressure. If the atmospheric pressure equals one atmosphere, boiling occurs at 100°C. Under normal conditions the barometric pressure varies from day to day and as a result the boiling point also varies slightly from 100°C. The barometric pressure decreases with increasing altitude and because of this, it may be necessary to cook foods for a longer period of time or use a pressure cooker to reduce cooking time. The lid on the pressure cooker forms a tight seal on the pot and is equipped with a pressure relief valve that prevents the pot from exploding. The heat supplied to the water causes the liquid water to evaporate until the pressure inside the cooker reaches the desired temperature and steam escapes via the pressure relief valve. Because the pressure inside the cooker is greater than one atmosphere, the water boils at a higher temperature and the food cooks faster.

Diffusion

Diffusion is the spontaneous movement of molecules from a region of high concentration to a region of low concentration due to random molecular movement. The spreading of a drop of food color in a glass of water and cigarette smoke in a room are examples of diffusion.

Diffusion occurs because molecules of all of the substances present are able to move throughout the container. However, in a liquid the molecules are much closer together than in a gas. Thus the molecules of one liquid diffusing into a second liquid undergo many more collisions per cm of distance than in the diffusion of one gas into another. The average distance between collisions, called the mean free path, is much shorter in a liquid. As a result, the rate of diffusion of one liquid into another is much slower for one liquid into another as compared to one gas into another.

Fick's Law of Diffusion

The rate of diffusion of one substance into another is described by the diffusion equation, or **Fick's law.**

$J = D\,A\,(C_1 - C_2)/d$

J is the rate of diffusion past a certain point each second and D is a constant of proportionality known as the diffusion constant or diffusion coefficient. D depends on the properties of the substances involved in the diffusion process. A is the cross-sectional area of the region through which the diffusion occurs.

C is the concentration of the diffusing substance in mol/m^3, where C_1 is the region of higher concentration and C_2 is the region of lower concentration. $(C_1 - C_2)/d$ is the concentration gradient. The concentration gradient is the change in concentration per unit distance (d).

The diffusion equation can also be written in terms of partial pressures:

$J_i = (D\,A/R\,T)\,(\Delta P_i\,/d)$

i is the particular component being considered, R is the universal gas constant, T is the temperature in K, and ΔP_i is the change in partial pressure over the distance (d) being considered.

The rate of diffusion of gas molecules can also be described by **Graham's law.** Graham's law states that the rate of diffusion of molecules (J) of a gas is inversely proportional to the square root of the molecular mass: $J \propto 1/(m)^{1/2}$.

EXAMPLE PROBLEM 6. Based on Table 13-4 in the textbook, the diffusion coefficient of blood hemoglobin through water is 6.90×10^{-11} m^2/s at 20.0°C and 1 atmosphere of pressure. Determine the rate of diffusion per m^2 of blood hemoglobin through water over a distance of 0.100 m from a region where the concentration is 3.0 mol/m^3 to a region where the concentration is 2.00 mol/m^3.

Part a. Step 1.	Solution: (Section 13-15)
Apply the formula for the rate of diffusion.	$J = D\,A\,(C_1 - C_2)/\Delta x$ and
	$J/A = D\,(C_1 - C_2)/\Delta x$
	$= [(6.90 \times 10^{-11}\ m^2/s)(3.00\ mol/m^3 - 2.00\ mol/m^3)]/(0.100\ m)$
	$J/A = 6.90 \times 10^{-10}\ mol/m^2\cdot s$

EXAMPLE PROBLEM 7. The rate of diffusion of an unknown gas (x) is determined to be 2.92 times faster than that of ammonia (NH_3). Determine the a) mass of the individual molecules of the unknown gas and b) molecular weight of the gas. The molecular weight of ammonia is 17.0 g/mol. Hint: use Graham's law of diffusion, which states that the rate of diffusion of a gas is inversely proportional to the square root of the molecular mass.

Part a. Step 1. Determine the mass of an ammonia molecule.	Solution: (Section 13-15) $m_{ammonia}$ = (17.0 g/mol)(1 mol/6.02 x 10^{23} molecules) $m_{ammonia}$ = 2.82 x 10^{-23} g/molecule
Part a. Step 2. Use Graham's law of diffusion to determine the mass of a molecule of the unknown gas.	From Graham's law of diffusion, $J \propto 1/(m)^{1/2}$; therefore, the ratio of the rates of diffusion of unknown gas to ammonia should be inversely related to the ratio of the square roots of their molecular masses, i.e., $(J_x)/(J_{ammonia}) = [1/(m_x)^{1/2}]/[1/(m_{ammonia})^{1/2}]$ but $J_x/J_{ammonia} = 2.92$ $2.92 = (m_{ammonia}/m_x)^{1/2}$ Using algebra to rearrange the equation: $m_x = (m_{ammonia})/(2.92)^2$ m_x = (2.82 x 10^{-23} g/molecule)/(8.53) = 3.31 x 10^{-24} g/molecule
Part b. Step 1. Determine the molecular weight of the unknown gas.	The molecular weight of a gas is the mass of 1 mole of the gas expressed in grams. The mass of 1 molecule is now known and 1 mole of a gas contains 6.02 x 10^{23} molecules. The molecular weight may now be determined. MW = (3.31 x 10^{-24} g/molecule) x (6.02 x 10^{23} molecules/mol) MW = 1.99 gram/mol The molecular weight of diatomic hydrogen (H_2) is 2.00 g/mole; therefore, the unknown gas is comprised of hydrogen molecules.

PROBLEM SOLVING SKILLS

For problems involving linear or volume expansion:

1. Complete a data table listing the initial length (or volume), coefficient of expansion, and change in temperature.
2. Determine the change in length (or volume). The final length (or volume) is the sum of the original plus the change.
3. If the problem involves a liquid expanding in a solid container, then the expansion of both the

liquid and the solid must be taken into account. The resulting coefficient of volume expansion equals the difference between the coefficient for the liquid and the coefficient for the solid.

For problems involving the ideal gas equation:

1. Complete a data table listing the absolute pressure, volume, temperature in degrees Kelvin, and the number of moles of gas present.
2. The absolute pressure equals the gauge pressure plus the atmospheric pressure.
3. Use the ideal gas equation to solve the problem.
4. If more than one gas is present, it may be necessary to determine the partial pressure exerted by each gas. The total pressure equals the sum of the partial pressures.

For problems involving the kinetic theory of gases:

1. Complete a data table listing the absolute pressure, volume, temperature in degrees Kelvin, and the number of moles of gas.
2. Solve for the average kinetic energy of the gas molecules.
2. Solve for the root-mean-square velocity of the molecules.

For problems involving rate of diffusion:

1. Complete a data table noting the concentration in the two regions, the distance over which the diffusion occurs, and the diffusion constant.
2. Use Fick's law to solve for the rate of diffusion.

For problems involving gaseous diffusion:

1. Complete a data table noting the molecular weight of each gas.
2. Use Graham's law to determine the rate of diffusion.

SOLUTIONS TO SELECTED TEXTBOOK PROBLEMS

TEXTBOOK PROBLEM 10. To make a secure fit, rivets that are larger than the rivet hole are often used and the rivet is cooled (usually in dry ice) before it is placed in the hole. A steel rivet 1.871 cm in diameter is to be placed in a hole 1.869 cm in diameter (at 20.0°C). To what temperature must the rivet be cooled if it is to fit in the hole?

Part a. Step 1

Assume that the final diameter of the rivet equals the diameter of the hole.

Solution: (Section 13-4)

$L_{final} = L_{original} + \Delta L = L_o + L_o\ \alpha\ \Delta T = L_o(1 + \alpha\ \Delta T)$

$1.869\ cm = (1.871\ cm)[1 + (12 \times 10^{-6}/°C)(T_f - 20°C)]$

$1.869\ cm = 1.871\ cm + (1.871\ cm)(12 \times 10^{-6}/°C)(T_f - 20°C)$

$-0.0020\ cm = (1.871\ cm)(12 \times 10^{-6}/°C)(T_f - 20.0°C)$

$-0.0020\ cm = (2.2 \times 10^{-5}\ cm/°C)(T_f - 20.0°C)$

$T_f - 20.0°C = (-0.0020\ cm)/(2.2 \times 10^{-5}\ cm/°C)$

$T_f - 20.0°C = -89°C$ and $T_f = -69°C$

TEXTBOOK PROBLEM 13. An ordinary glass is filled to the brim with 350.0 mL of water at 100°C. If the temperature decreased to 20°C, how much water could be added to the glass?

Part a. Step 1.

Determine the change in volume.

Solution: (Section 13-4)

Note: in determining the coefficient of volume expansion, both the water and the glass must be considered.

$\Delta V = V_o \, \beta \, \Delta T$

$\Delta V = (350.0 \text{ ml})(210 \times 10^{-6}/C^o - 27.0 \times 10^{-6}/C^o)(20^oC - 100^oC)$

$\Delta V = -5.1 \text{ mL}$

The negative value for the change in volume indicates that the water contracts more than the glass. Therefore, 5.1 mL of water is needed to fill the glass to the brim.

TEXTBOOK PROBLEM 21. (a) The tube of a mercury thermometer has an inside diameter of 0.140 mm. The bulb has a volume of 0.255 cm^3. How far will the thread of mercury move when the temperature changes from 11.5°C to 33.0°C? Take into account the expansion of Pyrex glass. (b) Determine a formula for the length of mercury column in terms of the relevant variables. Ignore the tube volume compared to the bulb volume.

Part a. Step 1.

Use Table 13-1 in the text to determine the coefficient of volume expansion for mercury and Pyrex glass.

Solution: (Section 13-4)

The coefficient of volume expansion of mercury is $180 \times 10^{-6}/C^o$ and $9.00 \times 10^{-6}/C^o$ for Pyrex glass.

Part a. Step 2.

Determine the change in volume.

Note: in determining the change in volume, the coefficient of volume expansion both the mercury and the Pyrex glass must be considered.

$\Delta V = V_o \, (\beta_{mercury} - \beta_{Pyrex}) \, \Delta T$

$\Delta V = (0.255 \text{ cm}^3)(180 \times 10^{-6}/C^o - 9.00 \times 10^{-6}/C^o)(33.0^oC - 11.5^oC)$

$\Delta V = 9.37 \times 10^{-4} \text{ cm}^3$

Part a. Step 3.

Determine the inside radius of the cylinder in cm.

radius $r = (½)$ diameter $= ½ \,(0.140 \text{ mm})(1.00 \text{ cm}/10.0 \text{ mm})$

$r = 7.00 \times 10^{-3} \text{ cm}$

Part a. Step 4. Determine the distance the mercury travels along the length of tube as a result of the expansion.	The mercury is expanding along the length of a cylindrical tube. The change in length (Δh) can be found as follows: $\Delta V = \pi\ r^2\ \Delta h$ $9.37 \times 10^{-4}\ cm^3 = \pi\ (7.00 \times 10^{-3}\ cm)^2\ \Delta h$ $\Delta h = 6.1\ cm$
Part b. Step 1. Derive a formula for the length of the mercury column in terms of the relevant variables.	from step 2: $\Delta V = V_o\ (\beta_{mercury} - \beta_{Pyrex})\ \Delta T$ from step 4: $\Delta V = \pi\ r^2\ \Delta h$ therefore, $\pi\ r^2\ \Delta h = V_o\ (\beta_{mercury} - \beta_{Pyrex})\ \Delta T$ $\Delta h = [V_o\ (\beta_{mercury} - \beta_{Pyrex})\ \Delta T]/[\pi\ r^2]$

Note: a fever thermometer has a constriction just above the bulb. When the thermometer is removed from the person's mouth the volume of the mercury tends to decrease because of the lower air temperature outside of the mouth. The thread of mercury is broken at the constriction before the mercury can begin to flow back into the bulb. Because of this, an accurate measurement of the person's temperature may be obtained. In order to repeat the reading it is necessary to shake the thermometer in such a manner that some of the mercury in the stem passes through the constriction and into the bulb.

TEXTBOOK PROBLEM 31. Calculate the density of oxygen at STP using the ideal gas law.

Part a. Step 1. Define what is meant by STP.	Solution: (Section 13-7 and 13-8) At STP, the temperature is 273 K and the pressure is 1.0 atm. Note: $1.0\ atm = 1.013 \times 10^5\ N/m^2$.
Part a. Step 2. Use the ideal gas law to determine the density. Note: 1.0 mol of O_2 = 32.0 g/mole.	$P\ V = n\ R\ T$ but $n = m/MW$ where m = the mass of the gas and MW = molecular weight of the gas $P\ V = (m/MW)\ R\ T$ $\rho = m/V = [P\ (MW)]/(R\ T)$ $\rho = [(1.013 \times 10^5\ N/m^2)(32.0\ g/mole)]/[(8.315\ J/mole\ K)(273\ K)]$ $\rho = 1430\ grams/m^3 = 1.43\ kg/m^3$

TEXTBOOK PROBLEM 47. Calculate the rms speed of helium atoms near the surface of the Sun at a temperature of about 6000 K.

Part a. Step 1. Determine the mass in kg of a helium atom.	Solution: (Section 13-10) m = (4.00 g/mole)(1.0 kg/1000 g)(1 mole/6.02 x 10^{23} atoms) m = 6.64 x 10^{-27} kg
Part a. Step 2. Calculate the rms speed of a helium atom.	$\overline{KE} = \frac{1}{2} m v^2$ but $\overline{KE} = 3/2\ k\ T$ $\frac{1}{2}(6.64 \times 10^{-27}\ kg)\ \overline{v^2} = (3/2)(1.38 \times 10^{-23}\ J/K)(6000\ K)$ $\overline{v^2} = 3.74 \times 10^7\ m^2/s^2$ $v_{rms} = 6 \times 10^3\ m/s$

TEXTBOOK PROBLEM 77. In outer space the density of matter is about 1 atom per cm^3, mainly hydrogen atoms, and the temperature is about 2.7 K. Calculate the rms speed of these hydrogen atoms, and the pressure (in atmospheres).

Part a. Step 1. Determine the mass in kg of a hydrogen atom.	Solution: (Section 13-10) m = (1.00 g/mole)(1.0 kg/1000 g)(1 mole/6.02 x 10^{23} atoms) m = 1.66 x 10^{-27} kg
Part a. Step 2. Calculate the rms speed of a helium atom.	$\overline{KE} = \frac{1}{2} m v^2$ but $\overline{KE} = (3/2)\ k\ T$ $\frac{1}{2}(1.66 \times 10^{-27}\ kg)\ \overline{v^2} = (3/2)(1.38 \times 10^{-23}\ J/K)(2.7\ K)$ $\overline{v^2} = 6.73 \times 10^4\ m^2/s^2$ $v_{rms} = 2.6 \times 10^2\ m/s$
Part a. Step 3. Determine the pressure in atmospheres. Note: 1 m^3 = 10^6 cm^3, 1 m^3 = 1000 liters, and 1 atm = 760 torr.	P V = n R T $P\ [(1\ cm^3)(1\ m^3/10^6\ cm^3)(1000\ liter)/(1\ m^3)] =$ $(1\ atom)[(1\ mol)/(6.02 \times 10^{23}\ atoms)](62.4\ liter\ torr/mol{\cdot}K)(2.7\ K)$ $P\ (10^{-3}\ liter) = 2.8 \times 10^{-22}$ liter torr $P = (2.8 \times 10^{-22}\ liter\ torr)/(10^{-3}\ liter) = 2.8 \times 10^{-19}$ torr $P = (2.8 \times 10^{-19}\ torr)/[(1\ atm)/(760\ torr)] \approx 4 \times 10^{-22}$ atm

CHAPTER 14

HEAT

OBJECTIVES

After studying the material of this chapter, the student should be able to:

- convert from joules to calories and kilocalories and vice versa.
- distinguish between the concepts of temperature and heat.
- explain what is meant by specific heat, latent heat of fusion, and latent heat of vaporization.
- apply the law of conservation of energy to problems involving calorimetry.
- distinguish the three ways that heat transfer occurs: conduction, convection, and radiation.
- solve problems involving the rate of heat transfer by convection and radiation.

KEY TERMS AND PHRASES

SI unit of heat is the joule (J). Also used are units related to the joule, namely, the **calorie** and **kilocalorie** where 4.184 J = 1 calorie (exactly) and 1000 cal = 1 kilocalorie.

internal energy consists of the total potential and kinetic energy of all of the atoms in the substance.

heat is the transfer of energy (usually thermal energy) from an object at higher temperature to one at a lower temperature.

specific heat is the amount of heat required to raise the temperature of 1 kg of a substance by 1°C. For water at 15°C, the specific heat is 1.00 kcal/kg°C or 4180 J/kg°C.

calorimetry is the quantitative measurement of the heat exchanged between two substances that are initially at different temperatures. Calorimetry is based on the law of conservation of energy and assumes that the heat lost by an object(s) at the higher temperature equals the heat gained by the object(s) at the lower temperature, i.e., heat lost + heat gained = 0.

change of phase refers to an object changing from the solid state to the liquid state and vice versa or from the liquid state to the gaseous state and vice versa.

latent heat of fusion refers to the energy that must be added or removed from a substance in order for the substance to change from the solid state to the liquid state or vice versa. The change occurs at the melting (freezing) point of the substance and no change of temperature occurs during the change of phase.

latent heat of vaporization refers to the energy that must be added or removed from a substance in order for the substance to change from the liquid state to the gaseous state or vice versa. The change occurs at the boiling (condensation) point of the substance and no change of temperature occurs during the change of phase.

conduction of heat is the result of collisions between molecules in a material. In a solid, the molecules are not free to move throughout the volume of the solid. The increase in kinetic energy of the molecules is in the form of vibrational kinetic energy.

convection is transfer of heat due to mass movement of warm material from one region to another. In the process of moving, the warm material displaces cold material.

radiation is the transfer of energy due to electromagnetic waves. Electromagnetic waves require no substance to transfer the energy.

thermal equilibrium between an object and its surroundings is reached when they reach the same temperature, i.e., $T_{1f} = T_{2f}$.

SUMMARY OF MATHEMATICAL FORMULAS

internal energy of an ideal gas	$U = 3/2\ n\ R\ T$	The internal energy (U) of an ideal gas depends on the number of moles (n) and the temperature in degrees Kelvin (T). R represents the universal gas constant.
heat transfer and specific heat	$Q = m\ c\ \Delta T$	The heat (Q) required to raise the temperature of a substance is related to the mass (m), specific heat (c) of the substance, and the change in temperature (ΔT).
latent heat of fusion or latent heat of vaporization	$Q = m\ \ell$	The heat (Q) added (or removed) in order to cause a substance to undergo a change of phase depends on the object's mass (m) and the latent heat (ℓ), where ℓ is either the heat of fusion (ℓ_F) or heat of vaporization (ℓ_V).
thermal conduction	$\Delta Q/\Delta t = k\ A\ (T_1 - T_2)/\ell$	The rate of heat flow ($\Delta Q/\Delta t$) through a substance due to conduction is related to the cross-sectional area of the object (A), the object's thickness (ℓ), and the temperature difference between the ends of the object ($T_1 - T_2$) in °C. k is the thermal conductivity of the material.

Stefan-Boltzmann's equation	$\Delta Q/\Delta t = e\ \sigma\ A\ T^4$	The rate at which an object radiates electromagnetic energy ($\Delta Q/\Delta t$) is related to the object's surface area(A), the object's temperature (T) in degrees Kelvin, and the emissivity (e) of the material. Stefan-Boltzmann's constant $\sigma = 5.67 \times 10^{-8}$ J/(s m^2 K^4).
net rate of radiant energy transferred	$\Delta Q/\Delta t = e\ \sigma\ A\ (T_1^4 - T_2^4)$	The net rate of radiant energy flow be tween an object and its surroundings is related to the surface area, emissivity and the difference in the temperatures between the object (T_1) and its surroundings (T_2).

CONCEPTS AND EQUATIONS

Units of Heat

The **SI** unit of heat is the joule (J). Also used are units related to the joule, namely, the **calorie** and **kilocalorie** where 4.184 J = 1 calorie (exactly) and 1000 cal = 1 kilocalorie. The kilocalorie is also known as the "large" calorie and is the unit associated with the energy content of foods.

Temperature, Heat, and Internal Energy

Temperature is the measure of the average kinetic energy of the individual molecules in a substance. All objects, whether solid, liquid, or gas, consist of atoms that are in motion. The **internal energy** consists of the total potential and kinetic energy of all of the atoms in the substance.

In an ideal monatomic gas we do not consider attractive or repulsive forces between atoms and so the internal energy (U) would be the total number of molecules in a gas times the average kinetic energy of the molecules.

$$U = N\,\overline{KE}$$

$$U = N\left(\tfrac{1}{2} m \overline{v^2}\right) = (3/2\ k\ T) = 3/2\ n\ R\ T$$

Heat is the transfer of energy (usually thermal energy) from an object at higher temperature to one at a lower temperature.

Specific Heat

The amount of heat (Q) required to raise the temperature of a substance depends on the quantity of matter or mass (m), the **specific heat** (c) of the substance involved and the change in temperature (ΔT). This is expressed mathematically as

$$Q = m\ c\ \Delta T$$

c is the specific heat of the substance. This is the amount of heat required to raise the temperature of 1 kg of a substance by 1°C. For water at 15°C, c = 1.00 kcal/kg°C = 4180 J/kg°C.

Calorimetry

Calorimetry is the quantitative measurement of the heat exchanged between two substances that are initially at different temperatures. For example, if hot water is added to a container of cold water, some of the heat energy of the hot water will be transferred to the cold water and also to the container. In calorimetry, this heat exchange is treated quantitatively.

Calorimetry is based on the law of conservation of energy and assumes that the heat lost by an object(s) at the higher temperature equals the heat gained by the object(s) at the lower temperature, i.e., heat lost + heat gained = 0 or heat gained = - heat lost.

EXAMPLE PROBLEM 1. A 0.0100 kg lead bullet moving at 100 m/s imbeds itself in a large block of wood. All of the kinetic energy lost by the lead bullet is transferred into heat which is shared equally by the bullet and the block of wood. Determine the temperature change of the bullet. The specific heat of lead is 0.0310 kcal/kg°C.

Part a. Step 1. Determine the amount of mechanical energy dissipated in the form of heat.	Solution: (Section 14-4) $Q = -\Delta KE = -(KE_f - KE_i)$ $= -(\frac{1}{2} m v_f^2 - \frac{1}{2} m v_i^2)$ $= -[\frac{1}{2}(0.0100\ kg)(0\ m/s)^2 - \frac{1}{2}(0.0100\ kg)(100\ m/s)^2]$ $Q = 50.0\ J$
Part a. Step 2. Determine the change in temperature of the bullet.	The heat transferred to the bullet $Q = \frac{1}{2}(50.0\ J) = 25.0\ J$ $Q = m\ c\ \Delta T$ $(25.0\ J)(1.0\ kcal/4180\ J) = (0.0100\ kg)(0.0310\ kcal/kg\ °C)\ \Delta T$ $\Delta T = 19.3\ C°$

Latent Heat

Latent Heat is the energy that must be added or removed from a substance in order for a **change of phase** to occur. For changes from the solid to the liquid state or vice versa the heat energy required is the **latent heat of fusion**. For changes from the liquid state to the gas state and vice versa the heat energy is the **latent heat of vaporization**.

Heat added to ice at 0°C causes ice to melt and a change of phase from the solid state to the liquid state occurs. There is no change in temperature during the phase change. The heat added overcomes the potential energy associated with the attractive forces between the molecules in the solid (ice). There is no increase in kinetic energy of the molecules until all of the ice melts and there is no increase in temperature during the phase change.

The reverse process occurs when water returns to the solid phase from the liquid phase. Heat must be removed from the system even though there is no change of temperature.

Table 14-3 in the textbook lists the latent heats of fusion and vaporization for a number of substances. For water, the heat of fusion is 79.7 kcal/kg or 333 kJ/kg and the heat of vaporization is 539 kcal/kg or 2260 kJ/kg. The heat required for a phase change is given by

$Q = m \ell$ where ℓ is the heat of fusion (ℓ_F) or heat of vaporization (ℓ_V).

TEXTBOOK QUESTION 6. Why does water in a canteen stay cooler if the cloth jacket surrounding the canteen is kept moist?

ANSWER: Latent heat is the energy that must be added or removed from a substance in order for a change of phase to occur. For changes from the liquid state to the gas state and vice versa the heat energy is the latent heat of vaporization.

Energy is required to cause water to evaporate from the cloth jacket. As the water evaporates, it removes energy from the metal container which makes up the shell of the canteen. As a result, the water in the canteen remains cooler than it would if the cloth jacket were not kept moist.

TEXTBOOK QUESTION 7. Explain why burns caused by steam on the skin are often more severe than burns caused by water at 100°C.

ANSWER: The heat of vaporization of water is 539 Kcal/kg and this heat is released when the steam strikes the skin and condenses into water.

If the heat present in 1 gram of steam as it condenses into water at 100°C is completely absorbed by the skin, the skin absorbs 539 cal of heat. If this same 1 gram of water now cools from 100°C to 37°C (body temperature), it releases only 73 cal of heat. The burns caused by steam at 100°C are more severe than an equal mass of water at 100°C.

TEXTBOOK QUESTION 9. Will potatoes cook faster if the water is boiling faster?

ANSWER: Under 1 atmosphere of pressure, water boils at approximately 100°C whether the water is boiling fast or slow. The time required for the potatoes to cook will be the same but energy will be saved if the water is set for a slow boil.

EXAMPLE PROBLEM 2. A 1.00 kg aluminum pot holds 2.00 kg of water at 20.0°C. Determine the amount of steam which must be added at 100°C to raise the temperature of the water and pot to 25.0°C. Note: the heat of vaporization of steam is 540 kcal/kg, while the specific heat of water is 1.00 kcal/kg °C, and the specific heat of aluminum is 0.220 kcal/kg °C.

Part a. Step 1. Determine the a temperature change for each substance in the system.	Solution: (Sections 14-4 and 14-5) Steam at 100°C to water at 100°C: no temperature change occurs during the change of phase to hot water. Hot water at 100°C to water at 25.0°C: $\Delta T_{hw} = 25.0°C - 100°C = -75.0\ C°$ Aluminum pot: $\Delta T_{Al} = 25.0°C - 20.0°C = 5.00\ C°$ Cold water at 20.0°C to water at 25.0°C: $\Delta T_{cw} = 25.0°C - 20.0°C = 5.00\ C°$
Part a. Step 2. Derive a formula for the heat lost by the steam and the hot water.	The steam becomes hot water which then loses heat until its temperature reaches 25.0°C. Therefore, the mass of the steam equals the mass of the hot water, i.e. $m_{hw} = m_{steam}$ heat lost by steam + hot water = $Q_{steam} + Q_{hw}$ $Q_{steam} + Q_{hw} = m_{steam}\,\ell_v + m_{hw}\,c_w\,\Delta T_{hw}$ $= m_{steam}(-540\text{ kcal/kg}) + m_{hw}(1.0\text{ kcal/kg°C})(-75.0°C)$ but $m_{steam} = m_{hw}$ $Q_{steam} + Q_{hw} = -[540\text{ kcal/kg} + 75.0\text{ kcal/kg}]\,m_{steam}$ heat lost = $-(615\text{ kcal/kg})\,m_{steam}$
Part a. Step 3. Determine the heat gained by the pot and cold water.	Note: let the mass of the cold water be m_{cw}. heat gained by pot + cold water = $m_{Al}\,c_{Al}\,\Delta T_{Al} + m_{cw}\,c_w\,\Delta T_{cw}$ $heat_{gained} = [(1.00\text{ kg})(0.220\text{ kcal/kg°C}) + (2.0\text{ kg})(1.0\text{ kcal/kg°C})](5.00°C)$ $heat_{gained} = 11.1\text{ kcal}$
Part a. Step 4. Apply the law of conservation of energy and solve for the mass of the steam.	The heat gained by the aluminum and cold water plus the heat lost by the steam and the hot water must equal zero. heat gained + heat lost = 0 $11.1\text{ kcal} + -(615\text{ kcal/kg})m_{steam} = 0$ $m_{steam} = (-11.1\text{ kcal})/(-615\text{ kcal/kg}) = 0.0180\text{ kg}$ or 18.0 grams

Heat Transfer: Conduction, Convection, and Radiation

There are three ways to transfer heat from one object or place to another. In a particular situation, one or more of the processes may be involved in the heat transfer.

Conduction

Conduction of heat is the result of collisions between molecules in a material. In a solid, the molecules are not free to move throughout the volume of the solid. The increase in kinetic energy of the molecules is in the form of vibrational kinetic energy.

If one end of the solid, e.g., a metal rod, is held in a flame, the molecules in that end will have a higher average vibrational kinetic energy than molecules further along the rod. The higher energy molecules transfer some of this energy to adjacent molecules via molecular collisions. As a result, heat energy is gradually transferred through the object.

The rate of heat flow through a substance due to conduction is given by the equation

$$\Delta Q/\Delta t = K\,A\,(T_1 - T_2)/\ell$$

$\Delta Q/\Delta t$ is the heat transferred per unit time, A is the cross-sectional area of the object, and ℓ is the object's thickness. $T_1 - T_2$ is the temperature difference between the ends of the object in °C, where T_1 is greater than T_2. K is the **thermal conductivity** of the material and the unit of K is kcal/(s m °C).

EXAMPLE PROBLEM 3. 400 ml of hot coffee at 60.0°C is placed in a closed styrofoam cup which is 1.50 mm thick and surface area 850 cm^2. The day is windy and the air outside the cup remains at a constant temperature of 20.0°C. The thermal conductivity of styrofoam is 5.5×10^{-6} kcal/s m°C. Determine a) the initial rate of heat flow through the styrofoam, and b) the time required for the temperature of the coffee to cool from 60.0°C to 50.0°C. Assume that 400 ml of coffee has a mass of 0.400 kg and that specific heat of coffee is the same as that of water.

Part a. Step 1.	Solution: (Section 14-6)
Complete a data table and apply the equation for the conduction of heat through a substance.	$\ell = (1.50\ \text{mm})(1\ \text{m}/1000\ \text{mm}) = 1.50 \times 10^{-3}\ \text{m} = 0.00150\ \text{m}$ $A = (850\ \text{cm}^2)(1\ \text{m}^2/10^{-4}\ \text{cm}^2) = 8.50 \times 10^{-2}\ \text{m}^2 = 0.0850\ \text{m}^2$ $T_1 - T_2 = 60.0°\text{C} - 20.0°\text{C} = 40.0\ \text{C}°$ $\Delta Q/\Delta t = k\,A\,(T_1 - T_2)/\ell$ $= (5.5 \times 10^{-6}\ \text{kcal/s m°C})(0.0850\ \text{m}^2)(40.0\ \text{C}°)/(0.00150\ \text{m})$ $\Delta Q/\Delta t = 1.25 \times 10^{-3}\ \text{kcal/s}$

Part b. Step 1.

Determine the time required for the temperature of the coffee to cool from 60.0°C to 50.0°C.

The time for the temperature of the coffee to change from 60.0°C to 50.0°C may be determined as follows:

heat loss = $(\Delta Q/\Delta t)$(time interval)

but heat loss = $Q = m\ c\ \Delta T$

$(\Delta Q/\Delta t)$(time interval) = $m\ c\ \Delta T$

$(1.25 \times 10^{-3}$ kcal/s$)\ t = -\,(0.400$ kg$)(1.0$ kcal/kg°C$)(50.0$°C $- 60.0$°C$)$

$t = 3200$ s or 53.3 minutes

Note: the answer is unrealistic due to a number of assumptions that would affect the time interval. For example, we have assumed that the cup is tightly sealed so that convective heat losses are negligible, rate of heat loss is constant, and the thickness of the cup is uniform. However, even with these assumptions we can see the advantage of using styrofoam cups for holding hot or cold liquids.

Convection

Convection is transfer of heat due to mass movement of warm material from one region to another. In the process of moving, the warm material displaces cold material. An example of this is a convection current present in a pan of water heated on a stove. The hot water rises from the bottom of the pan to the top while the cold water on the top sinks to the bottom. As the cold water sinks it is heated and the convection process continues until the boiling point is reached.

Radiation

Radiation is the transfer of energy due to electromagnetic waves. Electromagnetic waves, e.g. light from the sun, travel at the speed of light. They require no material medium to transport the waves, so light from very distant objects, e.g. the stars, is able to travel through the vacuum of outer space to the Earth. The rate at which an object radiates electromagnetic energy is given by the Stefan-Boltzmann's equation $\Delta Q/\Delta t = e\ \sigma\ A\ T^4$.

$\Delta Q/\Delta t$ is the rate at which energy leaves the object, A is the object's surface area, and T is the object's temperature in K. e is the **emissivity** of the material and is a characteristic property of the material. A perfect absorber is also a perfect emitter and has an emissivity value of 1. A perfect absorber is known as a perfect black body. A perfect reflector has an emissivity value of 0. $\sigma = 5.67 \times 10^{-8}$ J/(s m^2 K^4) and is known as the Stefan-Boltzmann constant. The net rate of radiant energy flow between an object and its surroundings is given by

$$\Delta Q/\Delta t = e\ \sigma\ A\ (T_1^4 - T_2^4)$$

T_1 and T_2 are the temperatures in degrees Kelvin of the object and its surroundings.

Thermal equilibrium ($\Delta Q/\Delta t = 0$) between an object and its surroundings is reached when they reach the same temperature, i.e., $T_1 = T_2$.

TEXTBOOK QUESTION 24. Why is the liner of a thermos bottle silvered (Fig. 14-15) and why does it have a vacuum between its two walls?

ANSWER: Silver reflects heat back into the liquid in the bottle and therefore heat loss due to radiation is kept to a minimum. The partial vacuum between the liner and the outer wall keeps heat loss due to convection to a minimum. The cork or styrofoam cap is a poor conductor of heat and if the seal is tight the rate of heat loss due to conduction is very low. Thus, the hot liquid in the thermos will be kept hot for several hours.

Note: the same arguments can be used to explain why cold liquids will remain cold for extended periods in a thermos.

EXAMPLE PROBLEM 4. A solid cylindrical metal bar is 2.00 cm in radius and is 20.0 cm long. The bar is heated until its temperature reaches 500°C. After it is removed from the source of heat it is placed in a room where the temperature is 20.0°C. Calculate the rate at which the bar radiates energy when it is first removed from the source of heat. Note: the emissivity value of the metal is 0.400.

	Solution: (Section 14-8)
Part a. Step 1. Calculate the temperature in degrees Kelvin.	273°C + 500°C = 773 K 273°C + 20.0°C = 293 K
Part a. Step 2. Determine the surface area of the bar.	The lateral surface area of the bar = $2\pi rL$ while the area of each end = πr^2, therefore $A = 2\pi\, r\, L + \pi r^2 + \pi r^2$ $A = 2\pi(2.00\text{ cm})(20.0\text{ cm}) + \pi(2.00\text{ cm})^2 + \pi(2.00\text{ cm})^2 = 276\text{ cm}^2$ $A = (276\text{ cm}^2)(1.0\text{ m}/100\text{ cm})^2 = 2.76 \times 10^{-2}\text{ m}^2$
Part a. Step 3. Determine the rate at which energy is radiated.	$\Delta Q/\Delta t = e\,\sigma\, A\, T^4$ $\Delta Q/\Delta t = (0.400)(5.67 \times 10^{-8}\text{ J/s m}^2\text{ K}^4)(2.76 \times 10^{-2}\text{ m}^2)[(773\text{ K})^4 - (293\text{ K})^4]$ $\Delta Q/\Delta t = 220\text{ J/s} = 220\text{ watts}$

PROBLEM SOLVING SKILLS

For problems involving calorimetry:

1. Complete a data table for information both given and implied in the problem. This information should list the initial and final temperature of each substance, the specific heat of each substance, and the mass of each substance. If a phase change occurs, the list should include the latent heat of fusion or vaporization of the substance involved in the phase change.

2. Apply the law of conservation of energy and solve the problem. Remember that the sum of the heat gain and loss equals zero.

For problems involving heat transfer by conduction:

1. Complete a data table listing information both given and implied as well as the unknown quantity. This information should include the temperature on either side of the material, the thickness of the material, the cross-sectional area through which the heat transfer occurs, and the thermal conductivity of the substance. Also, the heat transferred and the time required for the transfer to occur should be included in the list.
2. Apply the formula for the rate of heat transfer through a material and solve for the unknown quantity.

For problems involving heat transfer by radiation:

1. Complete a data table listing information both given and implied. This list should include the temperature of the object and its surroundings, the surface area of the object, the object's emissivity and the Stefan-Boltzmann constant. Also, the heat transferred and the time required for the transfer should be included in the list.
2. Apply the formula for rate of heat transfer by radiation and solve for the unknown quantity.

SOLUTIONS TO SELECTED TEXTBOOK PROBLEMS

TEXTBOOK PROBLEM 3. An average active person consumes about 2500 Cal a day. (a) What is this in joules? (b) What is this in kilowatt-hours? c) Your electric company charges about a dime per kilowatt-hour. How much would your energy cost per day if you bought it from the power company? Could you feed yourself on this much money per day?

Part a. Step 1. Convert from Calories to joules.	Solution: (Section 14-1) One food Calorie = 1000 calories = 4180 joules (2500 Cal)(4180 J/Cal) = 1.0×10^7 J
Part b. Step 1. Convert from joules to kWh.	1 kWh = 3.6×10^6 J (1.0×10^7 J)(1 kWh/3.6×10^6 J) = 2.92 kWh
Part b. Step 2. Determine the cost of this energy in cents.	(2.92 kWh)[(10 cents)/kWh] = 29 cents

Part b. Step 3. Could you feed yourself on this money?	The answer is no.

TEXTBOOK PROBLEM 11. A 35 g glass thermometer reads 21.6°C before it is placed in 135 ml of water. When the water and thermometer come to equilibrium, the thermometer reads 39.2°C. What was the original temperature of the water?

Part a. Step 1. Determine the change temperature for each substance.	Solution: (Section 14-4) glass: $\Delta T_g = 39.2°C - 21.6°C = 17.6°C$ water: $\Delta T_w = 39.2°C - T_i$
Part a. Step 2. Determine the mass of the water and glass in kg.	mass of the glass = (35 g)(1kg/1000 g) = 0.035 kg, mass of water = (135 ml)(1.0 g/ml)(1kg/1000 g) = 0.135 kg
Part a. Step 3. Apply the law of conservation of energy to determine the initial temperature of the water.	heat gained by glass + heat lost by water = 0 heat gained by glass = - heat lost by water $m_g\, c_g\, \Delta T_g = -\, m_w\, c_w\, \Delta T_w$ [(0.035 kg)(0.200 kcal/kg°C)(17.6°C) = - (0.135 kg)(1.00 kcal/kg°C)](39.2°C - T_i) 0.123 kcal = - 5.29 kcal + (0.135 kcal/°C)T_i T_i = (5.4 kcal)/(0.135 kcal/°C) = 40.1°C

TEXTBOOK PROBLEM 17. When a 290 g piece of iron at 180°C is placed in a 95 g aluminum calorimeter cup containing 250 g of glycerin at 10°C, the final temperature is observed to be 38°C. What is the specific heat of glycerin? Note: the specific heat of iron is 0.11 kcal/kg °C, and the specific heat of aluminum is 0.22 kcal/kg °C.

Part a. Step 1. Determine the heat lost by the iron.	Solution: (Sections 14-4 and 14-5) heat lost = $m_{Fe}\, c_{Fe}\, \Delta T_{Fe}$

$= (0.290 \text{ kg})(0.11 \text{ kcal/kg °C})(38°\text{C} - 180°\text{C})$

heat lost = - 4.5 kcal

Part a. Step 2.

Write an equation for the heat gained by the glycerin and the cup.

heat gained $= [m_{gly}\, c_{gly} + m_{Al}\, c_{Al}]\,(38°\text{C} - 10.0°\text{C})$

$= [(0.25 \text{ kg})\, c_{gly} + (0.095 \text{ kg})(0.22 \text{ kcal/kg°C})](28°\text{C})$

heat gained $= (7.0 \text{ kg °C})\, c_{gly} + 0.585 \text{ kcal}$

Part a. Step 3.

Apply the law of conservation of energy and solve for c_{gly}.

heat gained + heat lost = 0

$(7.0 \text{ kg °C})\, c_{gly} + 0.585 \text{ kcal} - 4.5 \text{ kcal} = 0$

$(7.0 \text{ kg °C})\, c_{gly} = 3.91 \text{ kcal}$

$c_{gly} = (3.91 \text{ kcal})/(7.0 \text{ kg °C})$

$c_{gly} = 0.56 \text{ kcal/kg °C} = 2.3 \times 10^3 \text{ J/kg °C}$

TEXTBOOK PROBLEM 26. An iron boiler of mass 230 kg contains 830 kg of water at 18°C. A heater supplies energy at a rate of 52,000 kJ/h. How long does it take for the water (a) to reach the boiling point, and (b) to all have changed to steam?

Part a. Step 1.

Use tables 14-1 and 14-3 in the textbook in order to complete a data table based on the information given.

Solution: (Section 14-5)

$m_{water} = 830 \text{ kg}$ $m_{iron} = 230 \text{ kg}$ $c_{water} = 4180 \text{ J/kg·°C}$

$c_{iron} = 450 \text{ J/kg·°C}$ $L_V = 22.6 \times 10^5 \text{ J/kg}$ (for water)

$\Delta T_w = \Delta T_{iron} = (100°\text{C} - 18°\text{C}) = 82 \text{ C°}$

Part a. Step 2.

Determine the amount of energy (Q_1) required to raise the temperature to the boiling point.

$Q_1 = Q_{water} + Q_{iron}$

$Q_1 = m_w\, c_w\, \Delta T_w + m_{iron}\, c_{iron}\, \Delta T_{iron}$ but $\Delta T_w = \Delta T_{iron}$

$= [(830 \text{ kg})(4180 \text{ J/kg·°C}) + (230 \text{ kg})(450 \text{ J/kg·°C})](100°\text{C} - 18°\text{C})$

$= [3.47 \times 10^6 \text{ J/°C} + 1.04 \times 10^5 \text{ J/°C}]\,(82 \text{ C°})$

$Q_1 = 2.93 \times 10^8 \text{ J}$

Part a. Step 3. Determine the time required for the water to reach the boiling point.	power = (energy required)/time $t_1 = Q_1/P$ where $P = 52{,}000$ kJ/h $= 5.2 \times 10^7$ J/h $= (2.93 \times 10^8 \text{ J})/(5.2 \times 10^7 \text{ J/h})$ $t_1 = 5.6$ h
Part b. Step 1. Determine the energy (Q_2) required to change the water at 100°C to steam at 100°C.	$Q_2 = m\,L_V$ $= (830 \text{ kg})(22.6 \times 10^5 \text{ J/kg})$ $Q_2 = 1.88 \times 10^9$ J
Part b. Step 2. Determine the time (t_2) required to change the water at 100°C to steam.	power = energy required/time $t_2 = Q_2/P$ $= (1.88 \times 10^9 \text{ J})/(5.2 \times 10^7 \text{ J/h})$ $t_2 = 36$ h
Part b. Step 3. Determine the total time.	$t_{total} = t_1 + t_2$ $= 5.6 \text{ h} + 36 \text{ h}$ $t_{total} \approx 42$ h

TEXTBOOK PROBLEM 37. Two rooms, each a cube 4.0 m on a side, share a 12 cm thick brick wall. Because of a number of 100 watt light bulbs in one room, the air is at 30°C, while the other room is at 10°C. How many of the 100 W light bulbs are needed to maintain the temperature difference across the wall?

Part a. Step 1. Complete a data table and apply the equation for the rate of heat transfer through a substance.	Solution: (Sections 14-6, 14-7 and 14-8) $\ell = (12 \text{ cm})(1 \text{ m}/100 \text{ cm}) = 0.12$ m $A = 4.0 \text{ m} \times 4.0 \text{ m} = 16 \text{ m}^2$ $\quad T_1 - T_2 = 30°\text{C} - 10°\text{C} = 20 \text{ C}°$ From Table 14-4 in the textbook, $k = 0.84$ J/s · m · C° $\Delta Q/\Delta t = k\,A\,(T_1 - T_2)/\ell$ $= (0.84 \text{ J/s}\cdot\text{m}\cdot\text{C}°)(16 \text{ m}^2)(20 \text{ C}°)/(0.12 \text{ m})$ $\Delta Q/\Delta t = 2240 \text{ J/s} = 2240$ W

Part a. Step 2. Determine the number of light bulbs required to maintain the temperature difference.	rate of heat transfer = number of light bulbs x power of each bulb 2240 W = n (100 W) n = 22.4 bulbs ≈ 23 light bulbs

TEXTBOOK PROBLEM 53. A mountain climber wears a goose down jacket 3.5 cm thick with a total surface area of 1.2 m^2. The temperature at the surface of the clothing is -20°C and at the skin is 34°C. Determine the rate of heat flow by conduction through the clothing (a) assuming that it is dry and the thermal conductivity, *k*, is that of down and (b) assuming the clothing is wet, so that *k* is that of water and the jacket is matted down to 0.50 cm thickness.

Part a. Step 1. Complete a data table and apply the equation for the conduction of heat through dry goose down.	Solution: (Section 14-6) $\ell = (3.5\ \text{cm})(1\ \text{m}/100\ \text{cm}) = 3.5 \times 10^{-2}\ \text{m} = 0.035\ \text{m}$ $A = 1.2\ \text{m}^2 \quad T_1 - T_2 = 34°\text{C} - -20°\text{C} = 54\ \text{C}°$ From Table 14-4 in the textbook, $k = 0.025\ \text{J/s} \cdot \text{m} \cdot \text{C}°$ $\Delta Q/\Delta t = k\,A\,(T_1 - T_2)/\ell$ $= (0.025\ \text{J/s} \cdot \text{m} \cdot \text{C}°)(1.2\ \text{m}^2)(54\ \text{C}°)/(0.035\ \text{m})$ $\Delta Q/\Delta t = 46\ \text{J/s} = 46\ \text{W}$
Part b. Step 1. Complete a new data table and apply the equation for the conduction of heat through wet goose down.	$\ell = (0.50\ \text{cm})(1\ \text{m}/100\ \text{cm}) = 0.0050\ \text{m}$ From Table 14-4 in the textbook, $k = 0.56\ \text{J/s} \cdot \text{m} \cdot \text{C}°$ $\Delta Q/\Delta t = k\,A\,(T_1 - T_2)/\ell$ $= (0.56\ \text{J/s} \cdot \text{m} \cdot \text{C}°)(1.2\ \text{m}^2)(54\ \text{C}°)/(0.0050\ \text{m})$ $\Delta Q/\Delta t = 7.3 \times 10^3\ \text{W}$

Note: based on the answer to part (b), the climber needs to change into dry clothing in a very short time or face the very real problem of losing his or her life due to hypothermia.

CHAPTER 15

THE LAWS OF THERMODYNAMICS

OBJECTIVES

After studying the material of this chapter, the student should be able to:

- explain what is meant by a physical system and distinguish between an open system and a closed system.
- state the first law of thermodynamics and use this law to solve problems.
- distinguish between an isothermal process, isobaric process, isochoric process, and adiabatic process and draw a PV diagram for each process.
- calculate the work done by a gas from a PV diagram. Use the equations for an ideal gas and for the internal energy of a gas to calculate the change in internal energy of a gas and the heat added or removed during a thermodynamic process.
- calculate the amount of heat which must be added or removed to change the temperature of a gas held in a closed container under conditions of constant volume or constant pressure.
- write from memory and explain the meaning of three equivalent ways of stating the second law of thermodynamics.
- use the first and second laws of thermodynamics to solve problems involving a Carnot engine.
- distinguish between a reversible process and an irreversible process. Give examples of each type of process.
- determine the change in entropy for a system in which the thermodynamic process is either reversible or irreversible.
- distinguish between macrostate and microstate and solve problems involving the statistical interpretation of entropy.

KEY TERMS AND PHRASES

thermodynamics is the study of energy transformations in natural processes and involves relations between heat, work, and energy.

system is any object or sets of objects which are under consideration; everything else in the universe is called the environment. In thermodynamics, a **closed** system is one where mass may not enter or leave. In an **open** system mass may be exchanged with the environment.

first law of thermodynamics is a statement of the law of conservation of energy. The first law states that the change in the internal energy (ΔU) of a closed system is due to heat added or removed from the system (Q) and/or work done on or by the system (W). $\Delta U = Q - W$.

pressure-volume diagram or PV diagram is used to determine the work done by a gas undergoing expansion or compression in a closed system.

isothermal process occurs when the temperature of the gas remains constant.

isobaric process occurs when the pressure is constant, the work done on or by the gas can be determined by using $W = P\,\Delta V$.

isochoric process occurs at constant volume, $\Delta V = 0$; therefore, $W = P\,\Delta V = 0$ and $Q = \Delta U$.

adiabatic process occurs when no heat flows into or out of the system. An adiabatic process usually occurs when a gas is compressed or expands so rapidly that there is no time for the heat to flow in or out of the system.

heat engine is a device that is capable of changing thermal energy (Q_H), also known as the input heat or heat of combustion of the fuel, into useful work (W).

Carnot engine is an idealized engine where energy losses due to internal friction, turbulence present in the fuel after ignition, etc., are not considered. Carnot determined that the maximum efficiency that can be realized from a heat engine depends on the temperature of the input heat and the exhaust heat.

refrigerators and air conditioners operate by removing heat from a low temperature (cold) reservoir and exhausting the heat to the higher temperature (hot) reservoir. In order to accomplish this task, work is done to cause heat to travel opposite from its normal direction.

entropy is a quantitative measure of the disorder in a system.

second law of thermodynamics can be stated in several equivalent ways, three of which are:
1) Heat energy flows spontaneously from a hot object to a cold object but not vice versa.
2) It is impossible to construct a heat engine which is 100% efficient. Thus a heat engine can convert some of the input heat into useful work, but the rest must be exhausted as waste heat.
3) The entropy of an isolated system never decreases. It can only stay the same or increase.

SUMMARY OF MATHEMATICAL FORMULAS

first law of thermodynamics	$\Delta U = Q - W$	The **first law of thermodynamics** is a statement of the law of conservation of energy. The first law states that the change in the internal energy (ΔU) of a closed system is due to heat added or removed from the system (Q) and/or work done on or by the system (W).

heat engine	$Q_H = W + Q_L$ or $W = Q_H - Q_L$	A heat engine is a device that converts thermal energy (Q_H) to useful work (W). The heat energy not converted to useful work is ejected to the environment as exhaust heat (Q_L).
maximum efficiency or Carnot efficiency of a heat engine	$e = W/Q_H$, $e = (Q_H - Q_L)/Q_H$ $e = (1 - Q_L/Q_H)$ $e = (T_H - T_L)/T_H$ $e = (1 - T_L/T_H)$	The maximum efficiency or Carnot efficiency (e) of a heat engine is equal to the ratio of the useful work (W) to the input heat (Q_H). The Carnot efficiency can be written in terms of the input temperature (T_H) and the exhaust temperature (T_L).
coefficient of performance (CP) for refrigerators or air conditioners	$CP = Q_L/W$ or $CP = (Q_L)/(Q_H - Q_L)$ $CP_{ideal} = (T_L)/(T_H - T_L)$	The CP for a refrigerator is the ratio of the heat removed from the cold region (Q_L) to the work (W) performed to remove the heat. The CP for an "ideal" refrigerator in terms of the temperatures of the low temperature reservoir (T_L) and the high temperature reservoir(T_H).
Entropy	$\Delta S = Q/T$	Entropy (S) is a quantitative measure of the disorder in a system. The change in entropy (ΔS) for a reversible process is directly proportional to the heat (Q) added to the system and inversely related to the temperature of the system (T). The unit of entropy is kcal/K, where the heat is measured in kcal and T in degrees Kelvin.
Statistical Interpretation of entropy	$S = k \ln W$ or $S = 2.3\ k \log W$	The entropy of the system (S) is propertonal to the number of ways or microstates that can occur. $k = 1.38 \times 10^{-23}$ J/K (Boltzsman's constant), ln is the logarithm to the base e, $e = 2.718$, log is the logarithm to the base 10, and W is the number of microstates corresponding to to the given macrostate.

CONCEPTS SUMMARY

Thermodynamics is the study of energy transformations in natural processes and involves relations between heat, work and energy. In this chapter we shall study the first and second laws of thermodynamics, their significance, and application.

Physical Systems

A **system** is any object or sets of objects which are under consideration, everything else in the universe is called the environment. In thermodynamics, a **closed** system is one where mass may not enter or leave. In an **open** system mass may be exchanged with the environment.

First Law of Thermodynamics

The **First Law of Thermodynamics** is a statement of the law of conservation of energy. The first law states that the change in the internal energy (ΔU) of a closed system is due to heat added or removed from the system (Q) and/or work done on or by the system (W).

$$\Delta U = Q - W$$

As a sign convention, heat added to a closed system is positive (Q = +), while heat removed is negative (Q = -). If work is done on the closed system, the internal energy and temperature increase and W is negative (W = -). If an ideal gas is compressed in a cylinder with a moveable piston, the temperature and internal energy increase. If the system does work on the surroundings (environment), the ideal gas pushes the piston outward and the gas expands, the internal energy and temperature decrease and W is positive (W = +).

PV Diagrams

The work done by a gas undergoing expansion or compression in a closed system can be determined through use of a **pressure-volume diagram** (PV diagram). The work done during an incremental volume change (ΔV) equals the area under the PV curve. This area may be determined using the process of graphical integration. The following figures represent typical PV processes on an ideal gas.

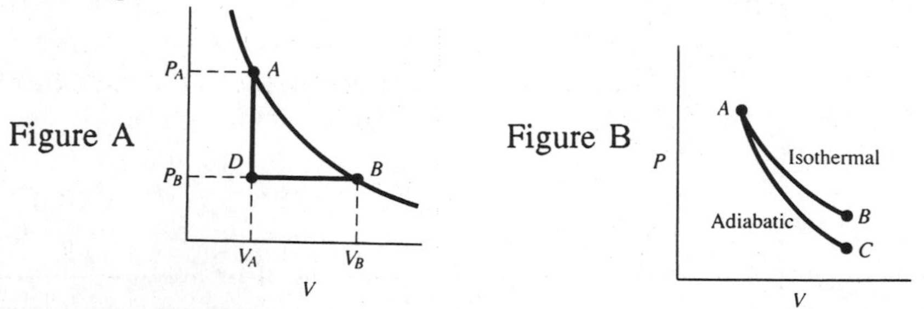

In an **isothermal** process, (AB) in Fig. A, the temperature of the gas remains constant, . $\Delta U = 0$, $Q = W$. Based on the general gas law, PV = nRT, if n and T are constant, then, PV = constant.

In an **isobaric** process, (DB) in Fig. A, the pressure is constant. The work done on or by the gas can be determined by using $W = P\,\Delta V$. n and P are constant; therefore V/T = constant.

An **isochoric** process, (AD) in Fig. A, occurs at constant volume, $\Delta V = 0$; therefore, $W = P\,\Delta V = 0$ and $Q = \Delta U$. For this process, P/T = constant.

In an **adiabatic** process, (AC) in Fig. B, no heat flows into or out of the system, Q = 0, and $W = -\Delta U$. An adiabatic process usually occurs when a gas is compressed or expands so rapidly that there is no time for the heat to flow in or out of the system. It should be noted that an adiabatic process is quite different from an isothermal process, although the PV diagram for each process appears similar.

TEXTBOOK QUESTION 3. In an isothermal process, 3700 J of work is done by an ideal gas. Is this enough information to tell how much heat has been added to the system? If so, how much?

ANSWER: The first law of thermodynamics can be written as $Q = \Delta U + W$. In an isothermal process the temperature remains constant. Since the temperature remains constant, the internal energy remains constant and $\Delta U = 0$. Therefore, all of the heat added during the process went into the work done by the ideal gas ($Q = W$). The amount of heat added to the system is 3700 J.

EXAMPLE PROBLEM 1. a) How much energy must be added to a 0.200 kg ice cube at 0.00°C in order to change it to water at 0.00°C. Determine b) the work done on the system during the change and c) the change in the internal energy of the system.

Part a. Step 1. Determine the energy required to melt the ice cube.	Solution: (Sections 15-1 and 15-2) The ice cube is undergoing a change of phase. Use the methods of Section 14-6 to solve for the energy required to melt the ice cube. $Q = m\,\ell_F$ $= (0.200\ \text{kg})(80\ \text{kcal/kg})(4180\ \text{J/kcal})$ $Q = 6.69 \times 10^4\ \text{J}$
Part b. Step 1. Determine the work done on the system during the change.	The change in volume of an ice cube when it melts is negligible. The external air pressure can be considered to be constant. Therefore, $W = P\,\Delta V$ $W = P\ (0\ \text{liters}) = 0\ \text{liter atm} = 0\ \text{joules}$
Part c. Step 1. Determine the change in the internal energy of the system.	The work done on the system is zero joules; therefore, all of the heat energy added increases the internal energy of the system. However, the temperature of the system does not change during the melting process and this indicates that the increase in the internal energy cannot be in the form of kinetic energy. The increase in the internal energy is in the form of potential energy as the molecules overcome the attractive forces which hold them in the solid phase. The increase in the internal energy may be determined as follows: $Q = \Delta U + W$ $6.69 \times 10^4\ \text{J} = \Delta U + 0\ \text{J}$ $\Delta U = 6.69 \times 10^4\ \text{J}$

EXAMPLE PROBLEM 2. One mole of an ideal gas is allowed to expand isothermally at -29.3°C from a volume of 4.0 liters and pressure 5.0 atm to a volume of 10.0 liters and pressure 2.0 atm. a) Draw a P-V diagram for the process, b) determine the work done by the gas during the expansion.

Part a. Step 1.

The process is isothermal, i.e., PV = constant. Complete a data table for P vs V with enough data points so that a reasonably accurate graph may be drawn.

Solution: (Section 15-2)

sample calculation

$$P_1 V_1 = P_2 V_2$$

$$(5.0 \text{ atm})(4.0 \text{ liters}) = (3.0 \text{ atm}) V_2$$

$$V_2 = 6.7 \text{ liters}$$

data table

P (atm)	V (liters)
5.0	4.0
4.0	5.0
3.0	6.7
2.0	10.0

Part a. Step 2.

Draw the PV diagram.

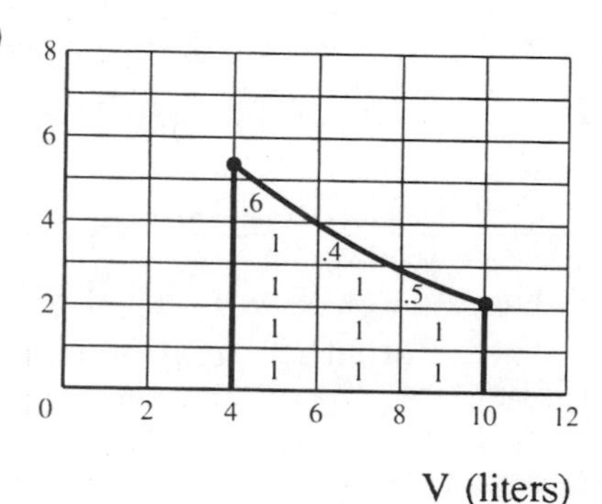

Part b. Step 1.

Determine the work done by the gas during the expansion.

The work done by the gas during the expansion is represented by the area under the curve between points A and B. The work done can be estimated by graphical integration, i.e., counting the complete and partial blocks under the curve and multiplying by the work represented by the area of 1 block. This method was previously used in chapter 2.

1.0 atm ▭ 2.0 liter

Work represented by one block = (1.0 atm)(2.0 liter) = 2.0 liter atm

sum of complete blocks = 9.0

sum of partial blocks = 0.4 + 0.6 + 0.5 + 0.1 = 1.6

Total number of blocks under the curve = 9.0 + 1.6 = 10.6

work done = (10.6 blocks)(2.0 liter atm/block) = 21.2 liter atm

Because of significant figures, the answer is 21 liter atm = 2100 J.

The Second Law of Thermodynamics

The **Second Law of Thermodynamics** can be stated in several equivalent ways, three of which are:

1) Heat energy flows spontaneously from a hot object to a cold object but not vice versa.
2) It is impossible to construct a **heat engine** which is 100% efficient. Thus, a heat engine can convert some of the input heat into useful work, but the rest must be exhausted as waste heat.
3) The **entropy** of an isolated system never decreases. It can only stay the same or increase. If the system is not isolated, then the change in entropy of the system (S_s) plus the change in entropy of the environment (S_{env}) must be greater than or equal to zero. The total entropy of any system plus that of its environment increases as a result of any natural process: $\Delta S = \Delta S_s + \Delta S_{env} > 0$.

First Statement of the Second Law

The first statement of the second law is a statement from common experience. When two objects, one hot and the other cold, come into contact, heat energy will be transferred from the system at higher temperature to the system at lower temperature, but not vice versa. While the first law states that energy must be conserved, i.e., the sum of the energy lost and gained in any process must equal zero, it does not say that heat must flow from the hot object to the cold object. The first law would not be violated if the hot object became hotter while the cold object became colder. The second law states that the direction of the heat flow must be from hot to cold.

Second Statement of the Second Law: Heat Engines

A **heat engine** is a device which is capable of changing thermal energy (Q_H), also known as the input heat or heat of combustion of the fuel, into useful work (W). Heat engines cannot be made to be 100% efficient and while part of the heat energy is converted to useful work, the remaining heat energy will be rejected to the environment or surroundings as waste heat (Q_L), e.g., exhaust from a car engine. Therefore,

$$Q_H = W + Q_L$$

In an idealized engine, known as a **Carnot engine**, energy losses due to internal friction, turbulence present in the fuel after ignition, etc. are not considered. Carnot determined that the maximum efficiency (e) that could be realized from a heat engine depends on the temperature of the input heat (T_H) and the exhaust heat (T_L), where T_H and T_L are expressed in degrees Kelvin.

The maximum efficiency or **Carnot efficiency** of a heat engine is determined by the following formulas:

$e = W/Q_H$, but since $W = Q_H - Q_L$, then

$e = (Q_H - Q_L)/(Q_H)$ and $e = (1 - Q_L/Q_H)$

Using the input and waste heat temperatures:

$e = (T_H - T_L)/(T_H)$ and $e = (1 - T_L/T_H)$

In order for a Carnot engine to be 100% efficient, it would be necessary for the temperature of the exhaust heat to be at absolute zero (zero degrees Kelvin). This is a practical as well as a theoretical impossibility.

EXAMPLE PROBLEM 3. A Carnot engine takes in 3000 calories of input heat and rejects 2000 calories as waste heat. The temperature of the waste heat is 600°C. Determine the a) efficiency of the engine and b) useful work done by the engine.

Part a. Step 1. Complete a data table based on information both given and implied in the problem.	Solution: (Section 15-5) Q_H = 3000 cal Q_L = 2000 cal T_H = ?	 T_L = (600 +273) K e = ? W = ?
Part a. Step 2. Determine the efficiency.	$e = (Q_H - Q_L)/(Q_H)$ = (3000 cal - 2000 cal)/(3000 cal) e = 0.333 or 33.3 %	
Part b. Step 1. Solve for the useful work done by the engine.	$e = W/Q_H$ 0.333 = W/3000 cal W = 1000 cal	alternate solution $W = Q_H - Q_L$ = 3000 cal - 2000 ca W = 1000 cal
Part c. Step 1. Solve for the temperature of the input heat	The efficiency and temperature of the waste heat are known, therefore, $e = (T_H - T_L)/(T_H)$ $0.333 = (T_H - 873\ K)/T_H$. T_H = 1310 K	

EXAMPLE PROBLEM 4. The actual work done by a heat engine in 1 hour is 2.7×10^6 J while frictional losses within the engine are 6.0×10^5 J. The engine operates between temperatures of 500 K and 1000 K. Determine the a) Carnot efficiency of the engine, b) number of joules of input heat taken in per hour by the engine and c) overall efficiency of the engine.

Part a. Step 1. Complete a data table using the information both given and implied in the problem.	Solution: (Section 15-5) Q_H = ? T_L = 500 K Q_L = ? e = ? T_H = 1000 K
Part a. Step 2. Determine the total work done by the machine.	$W_{total} = W_{actual} + W_{friction}$ $= 2.7 \times 10^6$ J + 6.0×10^5 J $W_{total} = 3.3 \times 10^6$ J
Part a. Step 3. Determine the Carnot efficiency of the engine.	Both the input heat temperature and waste heat temperature are known, therefore, $e = (T_H - T_L)/(T_H)$ = (1000 K - 500 K)/1000 K e = 0.50 or 50%
Part b. Step 1. Determine the input heat per hour.	In order to determine the input heat, it is necessary to use the total work output during 1 hour. $e = (W_{total})/(Q_H)$ $0.50 = (3.3 \times 10^6 \text{ J})/Q_H$ $Q_H = 6.6 \times 10^6$ J
Part c. Step 1. Determine the overall efficiency, i. e., the efficiency after frictional losses are taken into account.	The overall efficiency is related to the actual work done. $e = (W_{actual})/(Q_H)$ $= (2.7 \times 10^6 \text{ J})/(6.6 \times 10^6 \text{ J})$ e = 0.41 or 41%

Refrigerators and Air Conditioners

Refrigerators and air conditioners operate by removing heat from a low temperature (cold) reservoir and exhausting the heat to the higher temperature (hot) reservoir. In order to accomplish this task, work is done to cause heat to travel opposite from its normal direction.

The effectiveness of a particular refrigerator or air conditioner in accomplishing the removal of heat from the low temperature reservoir is measured by the coefficient of performance (CP). The CP for a refrigerator is the ratio of the heat removed from the cold region (Q_L) to the work

(W) performed to remove the heat, i.e.,

$CP = (Q_L)/(W)$ but $W = Q_H - Q_L$ so that $CP = (Q_L)/(Q_H - Q_L)$

The coefficient of performance for an "ideal" refrigerator can be written in terms of the temperature of the low temperature reservoir and the temperature of the high temperature reservoir as follows:

$CP_{ideal} = (T_L)/(T_H - T_L)$

TEXTBOOK QUESTION 7. Can you warm the kitchen in winter by leaving the oven door open? Can you cool the kitchen on a hot summer day by leaving the refrigerator door open?

ANSWER: Yes, you can warm the kitchen in winter by leaving the oven door open. In the case of an electric oven, electrical energy is converted to heat energy in the heating coils. The heating coils warm the air in the oven and the warm air will warm the room. If the oven is a gas oven, the gas flame heats the air and this warm air will warm the room.

You cannot cool the kitchen on a hot summer day by leaving the refrigerator door open. A refrigerator operates by removing heat from inside the refrigerator and releasing it into the air behind the refrigerator. If the door is left open, the air in front of the refrigerator would be cooled but the air behind the refrigerator would be warmed. The additional heat generated by the compressor would cause the overall temperature of the room to increase.

The author of this study guide remembers the house of a friend where the builder had cut a hole in the kitchen wall and inserted the refrigerator. The front of the refrigerator was in the kitchen while the back portion was in the garage. Therefore, the heat generated warmed the air in the garage while the air in the kitchen was cooled.

Reversible and Irreversible Processes

In an ideal gas, a reversible process is one in which the values of P, V, T, and U will have the same values in the reverse order if the process is returned to its original state. In this type of process, the ideal gas can be returned to is original state with no change in the magnitude of the work done or the heat exchanged. To be reversible, the process must be done very slowly, with no loss of energy due to dissipative forces such as friction and no heat conduction due to a temperature difference. In reality, these conditions cannot be met and all real processes are irreversible.

Third Statement of the Second Law: Entropy

Entropy (S) is a quantitative measure of the disorder in a system. The change in entropy (ΔS) for a reversible process is directly proportional to the heat (Q) added to the system and inversely related to the temperature of the system (T).

$\Delta S = Q/T$

The unit of entropy is kcal/K, where the heat is measured in kcal and T in degrees Kelvin.

The Kelvin temperature at which heat is added must remain constant for a process to be reversible.

As heat is added to a system, the average kinetic energy of the molecules increases and the motion becomes more disordered. At low temperatures we would expect to find a high degree of order while at high temperatures the system is likely to be very disordered. Thus, energy added while the system is at low temperature would introduce considerably more disorder than the same amount of energy introduced when the system is at a high temperature. Therefore, the change in entropy is inversely proportional to the temperature at which the heat is added.

Statistical Interpretation of Entropy

An equivalent definition of entropy can be given from a detailed analysis of the position and velocity (i.e. microstate) of every molecule which makes up the system. The entropy of the system is proportional to the number of ways that the microstates can occur and is given by the following formula:

$$S = k \ln W = 2.3\, k \log W$$

where k is Boltzmann's constant, $k = 1.38 \times 10^{-23}$ J/K. ln is the logarithm to the base e, where e = 2.718. log is the logarithm to the base 10. W is the number of microstates corresponding to the given macrostate.

The state of highest entropy is the state that can be achieved in the largest number of ways and is therefore the most probable.

TEXTBOOK QUESTION 16. (a) What happens if you remove the lid of a bottle containing chlorine gas? (b) Does the reverse process ever happen? Why or why not? (c) Can you think of other examples of irreversibility?

ANSWER: (a) Once the lid is removed, chlorine gas would gradually diffuse throughout the room. The gas molecules tend to move in random directions reaching a state of maximum disorder in agreement with the second law of thermodynamics. (b) Due to their random motions, a few molecules of chlorine gas will return to the bottle. However, the vast majority will not return to the bottle. This process is irreversible; the gas returning to the bottle would violate the second law of thermodynamics.

Another example of irreversibility would be placing 100 coins heads-up in a box and shaking the box vigorously. After shaken, there is only one way (chance) in 1×10^{29} that the coins could remain heads-up. The same odds apply for subsequent shaking of the box. A third example would be when a new deck of cards is opened the cards are in order according to their suit, i.e., hearts, diamonds, clubs or spades. Subsequent shuffling of the cards gives a random arrangement with high odds against ever finding the original arrangement again occurring.

TEXTBOOK QUESTION 19. Suppose you collect a lot of papers strewn all over the floor; then you stack them neatly. Does this violate the second law of thermodynamics? Explain.

ANSWER: Contrary to what teenagers tell their parents, putting the papers in order does not violate the second law of thermodynamics. An outside agent, possibly the teenager but probably the parent, does work in restoring order. There is a decrease of entropy for the papers but an overall increase in entropy due to the work done by the outside agent.

EXAMPLE PROBLEM 5. 100 coins are placed heads up in a box. After the box is shaken vigorously, only 50 coins are found to be heads up. Determine the change in entropy of the system.

Part a. Step 1.

Determine the entropy for each macrostate.

Solution: (Section 15-10)

Based on Table 15-3 in the textbook, there is only 1.0 microstate that corresponds to the macrostate where 100 heads are found. This is because in order to have 100 heads, each coin must come up heads. However, there are 1.0×10^{29} microstates which correspond to the macrostate where 50 coins come up heads. The entropy for each macrostate is determined from the formula $S = 2.3 \text{ k} \log W$.

1 head

$S = 2.3\,(1.38 \times 10^{-23} \text{ J/K})\,(\log 1.0)$

but $\log 1.0 = 0$; therefore, $S = 0$.

50 heads

$S = 2.3\,(1.38 \times 10^{-23} \text{ J/k})(\log 1.0 \times 10^{29})$

but $\log 1.0 \times 10^{29} = 29$

therefore, $S = 2.3\,(1.38 \text{ X } 10^{-23} \text{ J/K})(29)$

and $S = 9.2 \times 10^{-22} \text{ J/K}$

Part a. Step 2.

Determine the change in entropy of the universe for this process.

$\Delta S = S_{50 \text{ heads}} - S_{100 \text{ heads}}$

$= 9.22 \times 10^{-22} \text{ J/K} - 0 \text{ J/K}$

$\Delta S = +\,9.22 \times 10^{-22} \text{ J/K}$

The change in entropy is positive, the entropy of the universe has increased in the process.

PROBLEM SOLVING SKILLS

For problems involving PV diagrams:

1. Take note whether the process was isobaric, isothermic, isochoric or adiabatic.

2. Use the appropriate equation(s) to complete a data table of P vs V.
3. Use the data table to construct the PV diagram.
4. Use the technique of graphical integration to determine the work done during the process.
5. Use the ideal gas equation, the equation for the internal energy and the first law of thermodynamics to complete the solution of the problem.

For problems involving the Carnot engine:

1. Complete a data table listing the input and output heat, the input and output temperatures, the efficiency, and the useful work performed.
2. Use the equations involving Carnot efficiency and the first law of thermodynamics to solve the problem.

For problems related to change in entropy:

1. Determine the energy added or removed from the system.
2. Determine the temperature in degrees Kelvin at which the energy is added or removed.
3. Use $\Delta S = Q/T$ to determine the change in entropy.

For problems involving the statistical interpretation of entropy:

1. Determine the number of microstates corresponding to the given macrostate.
2. Use $S = k \ln W = 2.3\ k \log W$ to determine the entropy of the system.

SOLUTIONS TO SELECTED TEXTBOOK PROBLEMS

TEXTBOOK PROBLEM 5. A 1.0 L volume of air initially at 4.5 atm of (absolute) pressure is allowed to expand isothermally until the pressure is 1.0 atm. It is then compressed at constant pressure to its initial volume and lastly is brought back to its original pressure by heating at constant volume. Draw the process on a PV diagram, including numbers and labels for the axes.

Part a. Step 1.

Draw the PV diagram and complete a data table listing the information provided.

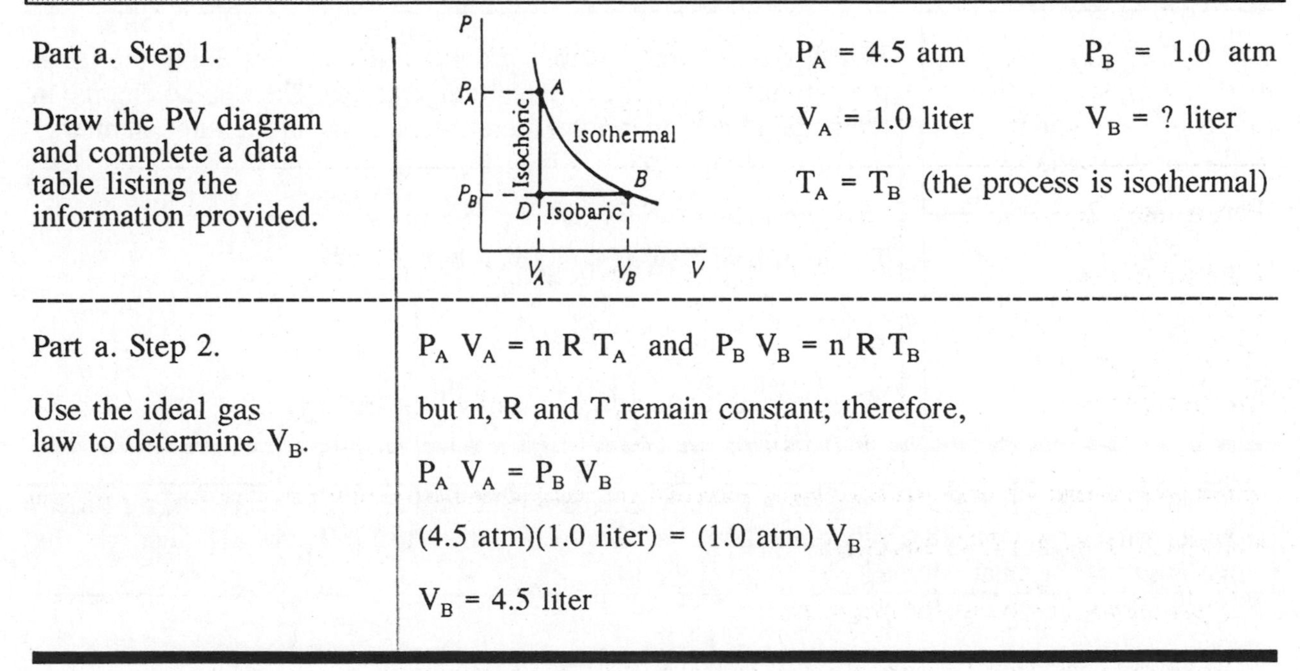

$P_A = 4.5$ atm $\qquad P_B = 1.0$ atm

$V_A = 1.0$ liter $\qquad V_B = ?$ liter

$T_A = T_B$ (the process is isothermal)

Part a. Step 2.

Use the ideal gas law to determine V_B.

$P_A V_A = n R T_A$ and $P_B V_B = n R T_B$

but n, R and T remain constant; therefore,

$P_A V_A = P_B V_B$

(4.5 atm)(1.0 liter) = (1.0 atm) V_B

$V_B = 4.5$ liter

TEXTBOOK PROBLEM 10. Consider the following two-step process. Heat is allowed to flow out of an ideal gas at constant volume so that its pressure drops from 2.2 atm to 1.4 atm. Then the gas expands at constant pressure, from a volume of 6.8 L to 9.3 L, where the temperature reaches its original value, see Fig. 15-22. Calculate (a) the total work done by the gas in the process, (b) the change in the internal energy of the gas in the process, and (c) the total heat flow into or out of the gas.

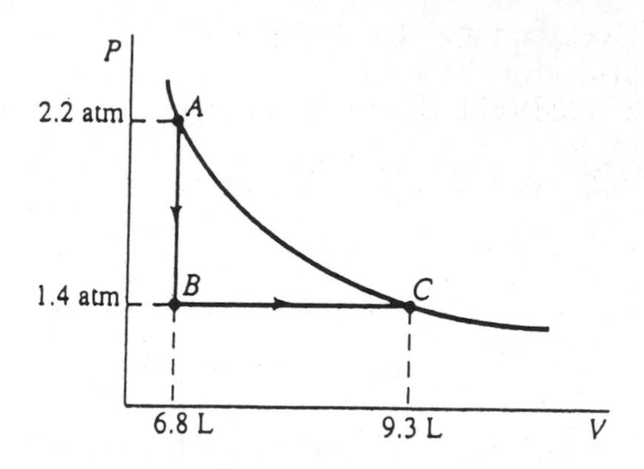

Part a. Step 1. Calculate the total work done by the gas.	Solution: (Sections 15-1 and 15-2) There is no change in volume between points A and B. Since $W = P\Delta V$, no work is done between A and B. $W_{total} = W_{AB} + W_{BC}$ where $W = P\,\Delta V$ $= 0\text{ J} + (1.4\text{ atm})(9.3\text{ L} - 6.8\text{ L})$ $W_{total} = (3.5\text{ liter atm})(101.3\text{ J/liter atm}) = 350\text{ J}$
Part b. Step 1. Determine the change in the internal energy.	Based on the information provided in the problem, the final temperature equals the original temperature. There is no change in temperature; and therefore, there is no change in the internal energy of the gas, i. e., $\Delta U = 0$ J.
Part c. Step 1. Determine the total heat flow into or out of the gas.	The total heat flow (Q) is given by $Q = W + \Delta U$ $Q = 350\text{ J} + 0\text{ J} = +350\text{ J}$ (into the gas)

TEXTBOOK PROBLEM 26. A Carnot engine utilizes a heat source at 550°C and has an ideal (Carnot) efficiency of 28%. To increase the ideal efficiency to 35%, what must be the temperature of the heat source?

Part a. Step 1. Convert the temperature of the heat source to Kelvin.	Solution: (Section 15-5) $T(K) = T(°C) + 273$ $T(K) = 550°C + 273°C = 823\ K$
Part a. Step 2. Use the equation for Carnot efficiency to solve for the exhaust temperature.	$e = (T_H - T_L)/T_H$ $0.28 = (823\ K - T_L)/(823\ K)$ $(0.28)(823\ K) = 823\ K - T_L$ $230\ K = 823\ K - T_L$ $T_L = 593\ K$
Part a. Step 3. Use the equation for Carnot efficiency to solve for the new temperature of the heat source.	$e = (T_H - T_L)/T_H$ where $e = 0.35$ $0.35 = (T_H - 593\ K)/T_H$ Note: T_L does not change $(0.35)\ T_H = T_H - 593\ K$ $-0.65\ T_H = -593\ K$ $T_H = 910\ K = 637°C$

TEXTBOOK PROBLEM 32. A heat pump is used to keep a house warm at 22°C. How much work is required of the pump to deliver 2800 J of heat into the house if the outdoor temperature is (a) 0°C, and (b) -15°C? Assume ideal (Carnot) behavior.

Part a. Step 1. Complete a data table.	Solution: (Section 15-6) $T_L = 0°C + 273°C = 273\ K$ $T_H = 22°C + 273°C = 295\ K$ $Q_H = 2800\ J$ $Q_L = ?$ $CP = ?$
Part a. Step 2. Determine the coefficient of performance.	Coefficient of performance (CP) for the heat pump is given by $CP = (T_H)/(T_H - T_L)$ $CP = (295\ K)/(295\ K - 273\ K)$ $CP = 13.4$
Part a. Step 3. Determine the work performed.	$CP = Q_H/W$ $13.4 = (2800\ J)/W$ $W = 210\ J$

Part b. Step 1. Complete a data table.	T_L = -15°C + 273°C = 258 K T_H = 22C + 273°C = 295 K Q_H = 2800 J Q_L = ? CP = ?
Part b. Step 2. Determine the coefficient of performance.	Coefficient of performance (CP) for the heat pump is given by $CP = (T_H)/(T_H - T_L)$ CP = (295 K)/(295 K - 258 K) = 7.97
Part b. Step 3. Determine the work performed.	$CP = Q_H/W$ 7.97 = (2800 J)/W W = 350 J

TEXTBOOK PROBLEM 35. What is the change in entropy of 250 g of steam at 100°C when it is condensed to water at 100°C?

Part a. Step 1. Determine the heat released by the steam as it changes to water.	$\Delta Q = m L_V$ Note: from Table 14-3, for water, $L_V = 22.6 \times 10^5$ J/kg $= (250 \text{ g})[(1 \text{ kg})/(1000 \text{ g})](-22.6 \times 10^5 \text{ J/kg})$ $\Delta Q = -5.65 \times 10^5$ J The answer is negative because heat is released by the steam.
Part a. Step 2. Determine the change in entropy.	$\Delta S = \Delta Q/T$ where T = 100°C = 373 K $= (-5.65 \times 10^5 \text{ J})/(373 \text{ K})$ $\Delta S = -1.51 \times 10^3$ J/K

TEXTBOOK PROBLEM 41. An aluminum rod conducts 7.50 cal/s from a heat source maintained at 240° C to a large body of water at 27°C. Determine the rate at which entropy changes per unit time in this process.

Part a. Step 1. Complete a data table.	Solution: (Section 15-7) T_{water} = 27°C + 273°C = 300 K T_{source} = 240°C + 273°C = 513 K Q/t = 7.50 cal/s = 31.4 J/s $(\Delta S_{total})/t$ = ?
Part a. Step 2. Determine the rate of change of entropy per unit time.	$(\Delta S_{total})/t = (\Delta S_{source})/t + (\Delta S_{water})/t$ but $\Delta S = Q/T$ $= (-31.4 \text{ J/s})/(513 \text{ K}) + (+31.4 \text{ J/s})/(300 \text{ K})$ $(\Delta S_{total})/t = -6.12 \times 10^{-2}$ J/s·K $+ 1.05 \times 10^{-1}$ J/s·K $= +4.35 \times 10^{-2}$ J/s·K